Introduction to the Numerical Analysis of Incompressible Viscous Flows

COMPUTATIONAL SCIENCE & ENGINEERING

Computational Science and Engineering (CS&E) is widely accepted, along with theory and experiment, as a crucial third mode of scientific investigation and engineering design. This series publishes research monographs, advanced undergraduate- and graduate-level textbooks, and other volumes of interest to a wide segment of the community of computational scientists and engineers. The series also includes volumes addressed to users of CS&E methods by targeting specific groups of professionals whose work relies extensively on computational science and engineering.

Series Volumes

Layton, William, *Introduction to the Numerical Analysis of Incompressible Viscous Flows*

Ascher, Uri M., *Numerical Methods for Evolutionary Differential Equations*

Zohdi, T. I., *An Introduction to Modeling and Simulation of Particulate Flows*

Biegler, Lorenz T., Omar Ghattas, Matthias Heinkenschloss, David Keyes, and Bart van Bloemen Waanders, Editors, *Real-Time PDE-Constrained Optimization*

Chen, Zhangxin, Guanren Huan, and Yuanle Ma, *Computational Methods for Multiphase Flows in Porous Media*

Shapira, Yair, *Solving PDEs in C++: Numerical Methods in a Unified Object-Oriented Approach*

Introduction to the Numerical Analysis of Incompressible Viscous Flows

William Layton
University of Pittsburgh
Pittsburgh, Pennsylvania

Society for Industrial and Applied Mathematics
Philadelphia

10 9 8 7 6 5 4 3 2 1

The cover was produced from images created by and used with permission of the Scientific Computing and Imaging (SCI) Institute, University of Utah; J. Bielak, D. O'Hallaron, L. Ramirez-Guzman, and T. Tu, Carnegie Mellon University; O. Ghattas, University of Texas at Austin; K. Ma and H. Yu, University of California, Davis; and Mark R. Petersen, Los Alamos National Laboratory. More information about the images is available at *http://www.siam.org/books/series/csecover.php*.

Library of Congress Cataloging-in-Publication Data

Layton, W. J. (William J.)
Introduction to the numerical analysis of incompressible viscous flows / William Layton.
p. cm. — (Computational science and engineering ; 6)
Includes bibliographical references and index.
ISBN 978-0-898716-57-3
1. Viscous flow—Mathematical models. 2. Numerical analysis. 3. Fluid mechanics. I. Title.
QA929.L39 2008
532'.0533015l8—dc22

2008018448

Contents

List of Figures ix

Foreword xi

Preface xiii

I Mathematical Foundations 1

1 Mathematical Preliminaries: Energy and Stress 3
1.1 Finite Kinetic Energy: The Hilbert Space $L^2(\Omega)$ 3
1.1.1 Other norms 7
1.2 Finite Stress: The Hilbert Space $X := H_0^1(\Omega)$ 8
1.2.1 Weak derivatives and some useful inequalities 10
1.3 Some Snapshots in the History of the Equations of Fluid Motion 12
1.4 Remarks on Chapter 1 15
1.5 Exercises 15

2 Approximating Scalars 17
2.1 Introduction to Finite Element Spaces 17
2.2 An Elliptic Boundary Value Problem 26
2.3 The Galerkin–Finite Element Method 30
2.4 Remarks on Chapter 2 33
2.5 Exercises 34

3 Vector and Tensor Analysis 37
3.1 Scalars, Vectors, and Tensors 37
3.2 Vector and Tensor Calculus 39
3.3 Conservation Laws 43
3.4 Remarks on Chapter 3 48
3.5 Exercises 49

II Steady Fluid Flow Phenomena **51**

4 Approximating Vector Functions **53**
4.1 Introduction to Mixed Methods for Creeping Flow 53
4.2 Variational Formulation of the Stokes Problem 56
4.3 The Galerkin Approximation 59
4.4 More About the Discrete Inf-Sup Condition 63
4.4.1 Other div-stable elements 66
4.5 Remarks on Chapter 4 66
4.6 Exercises 68

5 The Equations of Fluid Motion **71**
5.1 Conservation of Mass and Momentum 71
5.2 Stress and Strain in a Newtonian Fluid 74
5.2.1 More about internal forces 75
5.2.2 More about V 76
5.3 Boundary Conditions 78
5.4 The Reynolds Number 83
5.5 Boundary Layers 87
5.6 An Example of Fluid Motion: The Taylor Experiment 91
5.7 Remarks on Chapter 5 92
5.8 Exercises 95

6 The Steady Navier–Stokes Equations **99**
6.1 The Steady Navier–Stokes Equations 99
6.2 Uniqueness for Small Data 106
6.2.1 The Oseen problem 108
6.3 Existence of Steady Solutions 110
6.4 The Structure of Steady Solutions 114
6.5 Remarks on Chapter 6 117
6.6 Exercises 117

7 Approximating Steady Flows **121**
7.1 Formulation and Stability of the Approximation 121
7.2 A Simple Example 124
7.3 Errors in Approximations of Steady Flows 125
7.4 More on the Global Uniqueness Conditions 131
7.5 Remarks on Chapter 7 132
7.6 Exercises 133

III Time-Dependent Fluid Flow Phenomena **137**

8 The Time-Dependent Navier–Stokes Equations **139**
8.1 Introduction 139
8.2 Weak Solution of the NSE 141

8.3 Kinetic Energy and Energy Dissipation . . . 145
8.4 Remarks on Chapter 8 . . . 147
8.5 Exercises . . . 148

9 Approximating Time-Dependent Flows **151**
9.1 Introduction . . . 151
9.2 Stability and Convergence of the Semidiscrete Approximations . . . 154
9.3 Rates of Convergence . . . 158
9.4 Time-Stepping Schemes . . . 161
9.5 Convergence Analysis of the Trapezoid Rule . . . 165
9.5.1 Notation for the discrete time method . . . 165
9.5.2 Error analysis of the trapezoid rule . . . 168
9.6 Remarks on Chapter 9 . . . 175
9.7 Exercises . . . 176

10 Models of Turbulent Flow **179**
10.1 Introduction to Turbulence . . . 179
10.2 The K41 Theory of Homogeneous, Isotropic Turbulence . . . 181
10.2.1 Fourier series . . . 182
10.2.2 The inertial range . . . 183
10.3 Models in Large Eddy Simulation . . . 186
10.3.1 A first choice of ν_T . . . 189
10.4 The Smagorinsky Model for ν_T . . . 190
10.5 Near Wall Models: Boundary Conditions for the Large Eddies . . . 192
10.6 Remarks on Chapter 10 . . . 194
10.7 Exercises . . . 195

Appendix Nomenclature **197**
A.1 Vectors and Tensors . . . 197
A.2 Fluid Variables . . . 197
A.3 Basic Function Spaces and Norms . . . 198
A.3.1 Other norms . . . 198
A.4 Velocity and Pressure Spaces and Norms . . . 199
A.5 Finite Element Notation . . . 200
A.6 Turbulence . . . 200

Bibliography **203**

Index **211**

List of Figures

1.1 This flow is exciting but far beyond what is reliably computable! 4

2.1 A curve and its piecewise linear interpolant. 18
2.2 A typical basis function. 19
2.3 Mesh with two "bad" triangles (upper left and lower right). 20
2.4 Mesh following a curve. 20
2.5 Mesh for flow around cylinder at Re = 40. 21
2.6 Different mesh density for Re = 200. 21
2.7 The finite element space X^h. 22
2.8 A sketch of the basis function $\phi_j(x, y)$. 22
2.9 The correspondence between basis functions and nodes. 23
2.10 A mesh resolving a circular transition region. 24
2.11 One of the three linear basis functions. 25
2.12 The cubic bubble function. 25
2.13 An inhomogeneous medium. 27
2.14 A typical adaptive FEM mesh. 33

3.1 Geometry of a simple shear flow. 41

4.1 Typical experimental realization of creeping flow. 54
4.2 Velocity vectors for Stokes flow. 55
4.3 Streamlines for Stokes flow. 55
4.4 Linear-constant pair violates stability. 60
4.5 The MINI element. 65
4.6 Another element satisfying the discrete inf-sup condition. 69

5.1 Stress-deformation relation is nonlinear. 77
5.2 Verifying no-slip at low stresses. 79
5.3 Driven cavity domain and boundary conditions. 80
5.4 An example of flow in the driven cavity. 80
5.5 Discontinuous boundary velocities induce infinite stress. 81
5.6 A fluid flows across a surface. 82
5.7 A typical flow over a step. 82
5.8 Exploring the Cauchy stress vector. 83

5.9 Geometry of the forward-backward step. 83
5.10 Flow in pipes at increasing Reynolds numbers. 84
5.11 A boundary layer flow. 88
5.12 Depiction of a boundary layer. 89
5.13 Setup of counter rotating cylinders. 92
5.14 Three Taylor cells. 93
5.15 Schematic of flow between rotating cylinders. 94

6.1 Velocity vectors for Re = 1. 100
6.2 Streamlines for Re = 1. 100
6.3 Velocity vectors for Re = 40. 100
6.4 Streamlines for Re = 40. 100
6.5 Velocity vectors Re = 200. 101
6.6 Vorticity contours Re = 200: the von Karman vortex street. 101
6.7 Vorticity contours at Re = 1000. 102
6.8 Voticity contours at Re = 2000. 102
6.9 FEM mesh for Re = 40. 103
6.10 FEM mesh for Re = 200. 103
6.11 Shear flow between parallel plates. 111
6.12 The map T(N(u)). 113
6.13 Behavior of flow between rotating cylinders. 115

8.1 Depiction of k and epsilon. 147

9.1 Streamtubes in a “simple” three-dimensional flow. 152

10.1 A Gaussian filter (heavy) and rescaled (thin). 187
10.2 A curve, its mean (heavy line) and fluctuation (dashed). 188
10.3 Eddies are shed and roll down channel. 191
10.4 Smagorinsky model predicts flow reaches equilibrium quickly. 192
10.5 $\bar{u}$ does *not* vanish on $\partial\Omega$. 193

Foreword

The flows of liquids and gases have forever fascinated man. They appear as the Hellenic myth of Charybdis and in the biblical tale of the parting of the Red Sea. They have attracted the interest of some of the great geniuses of physics, engineering, and mathematics, including Leonardo, Euler, Cauchy, and Prandtl. They were why electronic computers were invented and remain the main driving force behind the development of today's supercomputers.

Given the long history of scientific inquiry into fluid flows and the great minds that have participated in the making of that history, one might be tempted to conclude that fluid mechanics is a stale, well-worn, and exhausted subject. Nothing could be further from the truth. Fluid mechanics remains a vital, viable, and vibrant area of research in mathematics, computations, modeling, and technological applications. Consider some examples. Existence and uniqueness questions related to the Navier–Stokes equations constitute one of the Clay Institute's million-dollar-prize problems. The efficient, accurate, and robust computation of turbulent flows remains an unresolved mystery. Bio- and nanofluidics present some very interesting and immediate modeling challenges. Industries of all types, e.g., aerospace, automotive, chemical, environmental, petroleum, pharmaceutical, and transportation, are all highly dependent on accurate simulations of fluid flows.

One should not be surprised to learn that hundreds of books have been written devoted to the subject of fluid mechanics, including many on computational aspects. One may naturally ask, why then another book? Bill Layton's book provides proof that another book can be and is of interest. In a very concise and efficient way, it provides a fresh perspective to well-studied subjects. It combines the well known with the recently discovered to effect the new perspectives. As such, the book is of interest to experts as well as novices and provides not only interesting reading but also valuable and useful information to practitioners.

Max Gunzburger
Tallahassee, FL
October 2006

Preface

But we are all led and guided by the passion to perceive and to understand,. . . .
L. Euler, from the preface to: *Considerations on Nautical Problems.*

The accurate, efficient, and reliable simulation of problems involving the flows of liquids and gasses is necessary for scientific and technological progress in many areas. In fact, decisions which affect our everyday lives are made daily based upon computational simulations which are often performed for flow problems far beyond those which can be reliably computed! Computational fluid dynamics (CFD) is also an area in which appetite for computational resources has always exceeded supply and will continue to do so for the foreseeable future.

CFD is one of the current and central scientific frontiers, and there are still important contributions which will be made by mathematicians in this area. However, this is an area not easily accessible to mathematics students. Before reading current papers in the area, students need to learn analysis, functional analysis, partial differential equations, numerical analysis of partial differential equations, continuum mechanics, mathematical fluid mechanics, and so on. On top of this, they are expected to develop some physical understanding and insight into the physics of fluids.

The first known mathematical fluid mechanics book is *Hydrostatics* by the great Archimedes. Throughout history, mathematicians have been key contributors to the development of the understanding of fluid motion. Yet, in the current training of mathematics students, a few basic courses are taken, after which they work exclusively with one professor. It is little wonder that with each generation, mathematical researchers become more specialized and narrow! This natural progression makes it more difficult for each succeeding generation to reach the real scientific frontier of an area like CFD. Progress in CFD requires communication between experts in numerical analysis, fluid dynamics, and large-scale computing with constant comparison against the behavior of real fluids in motion.

This book was written to help graduate students who feel they are up to the challenge of the beautiful and complex world of CFD. The purpose of this book is to allow graduate students to progress from essentially zero to finite element CFD and even include one advanced topic in the field (such as turbulence[1]) in one academic term. Because this book has been written for graduate students, there is a lot of repetition in the presentation. The

[1]Turbulence is the example included as Chapter 10. There are many other important applications which, depending on the interests of the instructor, can be presented instead. Chapters 1 through 9 are designed to give students the foundation for many of them.

focus of this book is on incompressible viscous flows. This is the case best understood mathematically (and for which central issues are still unresolved). There are many other fluid flow problems whose extra difficulties build upon those of incompressible viscous flows, such as viscoelasticity, plasmas, compressible flows, coating flows, flows of mixtures of fluids, and bubbly flows. The world of fluid motion is fantastically varied and complex. A good understanding of the interconnections among the physics, the mathematics, and the numerics of the incompressible case is valuable, possibly even essential, for progress in these more complex flows.

For this purpose, this book must pick a path through finite element CFD which is both mathematically cogent and physically lucid. The path this book takes is energy (in)equality. The energy equality for a viscous incompressible fluid is a mathematical estimate fairly easily derived from the system of partial differential equations. More than that, it is the direct link between the Navier–Stokes equations (NSE) and the fundamental physics of fluid motion, stating in precise mathematical terms that

> *kinetic energy at time t + total energy dissipated up to time t balances initial kinetic energy + total kinetic energy input up to time t.*

This energy equality implies *stability*[2] of the velocity: various norms of the fluid velocity are bounded by other norms of the problem data. Working backward from this fundamental physical fact and mathematical theorem, this book presents the mechanics of the equations of motion, their mathematical architecture, and the necessary analytical background. Working forward from the energy estimate, the book traces the energy norm path through the stability of finite element methods (FEMs) and their error analysis. In the last chapter, the K41 theory of turbulence is presented as a simple outgrowth of the energy equality. At the end, readers will have an intuitive yet mathematically rigorous connection from beginning to end. This thread might be thin after only one term, but students are then prepared to read (better) books and articles on the specific topics of the chapters and integrate what they have read within the overall picture of finite element CFD.

There are many challenges facing students beginning this study. It is always preferable to have a full and complete background before entry. While this is possible for some students, most find their background less than ideal. (Researchers in the area find it necessary to keep learning, too.) Many students in our program are still learning real analysis when they begin this book.

Several choices had to be made for this book. The first was to focus the proofs in the book purposely on the *proofs* the students must really master at their *first* entry to finite element CFD. Thus, when a result is important but its *proof* is not central, the result is quoted and the proof referenced to other books. The central proofs are presented with redundancy and extra explanations that experts could find tedious. The second choice regards the assumed analysis background of the student. These notes have tried to minimize this as far as possible, consistent with correctness and relevance. Chapter 1 begins with the unavoidable chapter on mathematical preliminaries. One choice in teaching a course based on this book is to postpone the material in this chapter (or at least the more technical

[2]There are many different types of stability of fluid motion, reflecting the needs of the variety of important applications of the area. Stability here is used in the simplest sense: the solution is bounded by the data of the problem.

subsections 1.1 and 2.1) until it is used and then introduce it bit by bit. (This is what I do.) The third choice was to write a book for *students* entering the field, assuming as little background as possible. This forces many interesting and important topics, such as duality, a posteriori error estimation, and adaptivity, to be left for a second treatment or further reading. Many students in our program have read (an early version of) this book on their own with minimal help and have gone on to do interesting work on many of these other topics.

The hardest choice about topics concerns what to do about implementation. This most important topic has been omitted for several reasons. The first one is simply time: an introduction (providing the foundation for reading, contributing as a mathematician, and communicating with other specialists) should be covered in one term. Second, the technical details about programming the methods become interesting mainly after seeing that the methods work (and work well) so extensions are needed. To see that the methods work well, there are excellent and easy-to-use programs available for finite element CFD; for example, COMSOL Multiphysics (previously FEMLAB) and FREEFEM++ are very friendly to students. In going from two to three space dimensions, real difficulties arise which require parallel codes. The code ViTLES, developed by Traian Iliescu and Jeff Borggaard, is a parallel, three-dimensional, NSE platform for both laminar and turbulent flows. My friends Vince Ervin and John Burkardt (Flow7) both have elegant finite element CFD programs on their web pages that my students have benefitted from.[3] The third reason for not presenting implementation at length herein is that implementation of the FEM for the NSE certainly deserves a book of its own!

I have taught a course based on this book with success at the University of Pittsburgh. This class included many beginning graduate students with interest (but not background) in analysis and applied mathematics and some more advanced graduate students already doing research in other areas of computational mathematics. For almost all of these students, this course was their first exposure to fluids, and for many of the students, the course was their first exposure to analysis beyond that of a typical beginning graduate student. In addition, I have had many students from engineering and physics who have enjoyed the mathematical presentation of CFD. All these students were users of CFD technology in their research and were led to seek a more systematic understanding of *why*, *when*, and *how* typical algorithms (do and don't) work. This book is far from appropriate for the first exposure of an engineering student (for example) to CFD but an excellent later course.

For success, I found three factors essential. The first is the commitment of students. They should understand that they will be learning a lot and that in true learning one is never in one's comfort zone. The second factor is active learning: students should try to work at least one exercise regularly after each section or lecture. (I include a few carefully chosen exercises in the text.) The third factor is a glimpse of the broader world of computational fluid dynamics beyond these notes. This can be done in many ways to fit the interests of the students and teachers. One way that is exciting for everyone is to have groups explore some of the available FEM CFD programs and report periodically.

Welcome to the exciting and beautiful world of fluids in motion and finite element computational fluid dynamics!

---

[3] www.freefem.org, www.icam.vt.edu/ViTLES, www.csit.fsu.edu/~burkardt/f_src/flow7/flow7.html

Overview of the Topics

Two fundamental mathematical arguments that provide support for finite element approximations of the NSE are

1. convergence of the approximate velocity and pressure of the steady NSE for laminar flows (i.e., at small enough Reynolds number to ensure global uniqueness) and
2. convergence in the energy norm of the semidiscrete approximation to the solution of the time-dependent problem.

Much research on the numerical analysis of the NSE is a reaction to the limitations of these two arguments or an elaboration of them to new problems and algorithms. These convergence results are themselves extensions of three (far more important) stability results:

1. The approximate pressure of the Stokes problem is bounded by problem data under the discrete inf-sup condition.
2. The approximate velocity of the steady problem is bounded by problem data if the nonlinearity is explicitly skew-symmetrized.
3. The kinetic energy in the approximate velocity of the time-dependent problem is bounded by problem data with the same treatment of the nonlinearity.

These three evolved from the corresponding energy estimates for the NSE which express a fundamental and direct link to the physics of fluid motion. Working backward to the preparation of a typical student entering graduate school in the mathematical sciences, focusing on what convergence analysis can tell about fluid flow phenomena (and what it leaves unsaid as well), the topics and order in this book emerged. The book's goal is to present a connected thread of ideas in the numerical analysis of the NSE without losing sight of understanding what fluid flow simulations really mean. Since the background of students is highly variable, each chapter is also presented to be as self-contained as possible so that students can begin at their appropriate place in the book. This means that essential definitions and results often reappear in chapters after they are first introduced.

Inevitably, this book begins with a chapter called Mathematical Preliminaries. Such a chapter must be there for a book to be read by a student without a teacher to supply gaps in the student's background. There are the usual choices: cover the chapter in detail, skip this chapter, totally introducing the topics in Chapter 1 as they are needed, or just introduce $L^2(\Omega)$ and $H_0^1(\Omega)$ and move ahead quickly, introducing the other topics and results as needed. (This is what I do.) Students are often anxious to see methods and applications before believing that theory is justified. Chapter 2 introduces (as quickly and as cleanly as possible) the FEM in enough detail to begin the treatment of essential ideas in the NSE. Many students (but not all) will have had some experience in the FEM. Chapter 2 is for students who are new to the FEM (and is not a substitute for a course on the topic). It presents energy norm stability and convergence of the FEM for the linear element and small extensions. The extensive, important, and intricate theory of the FEM is motivated by exactly this case, so this is the place to start and it is enough. Chapter 2 does not take the path (traditional since the finite element book of Strang and Fix) of beginning in one dimension then repeating the presentation in two dimensions. This approach (which is best for a full term course on FEMs) takes a lot of time. I have found the key to students understanding the FEM at some intuitive level is pictures, drawings, schematics, etc., of meshes and basis functions, all possible for linear elements on triangles in two dimensions. Thus, the FEM chapter begins in two dimensions.

Engineering and science students are (in my experience) adept at calculations with vectors and tensors, but, alas, these topics are treated only very briefly in some undergraduate mathematics curriculums.[4] Thus, Chapter 3 covers and reviews vectors, tensors, and conservation laws. It presents only that which is necessary to go farther in CFD.

Part II opens with Chapter 4, which presents mixed methods for the Stokes problem. The message of Chapter 4 is centrality of stability of the discrete pressure. This leads to the continuous inf-sup condition and its discrete analogue. Every chapter in this book has multiple excellent books on the chapter's topic, and Chapter 4 is no exception. By emphasizing stability of the discrete pressure, the treatment of mixed methods is shortened and much of the important and beautiful theory of mixed methods is left for the students' next steps in the field. Still, a clear understanding of

$$LBB^h \Rightarrow \text{stability of } p^h \Rightarrow \text{ convergence}$$

is essential for reading and understanding the theory of mixed methods. It may seem odd to place the Stokes problem before the derivation of the equations of fluid motion. This was done to keep the students focused on numerics. Each theory chapter is followed by a numerical analysis chapter connected to and expanding the abstract theory. The numerical analysis of the Stokes problem can be presented before the derivation of the NSE, and to delay discussion of numerical issues longer risks losing the interest of many students.

Chapter 5 gives a derivation of the NSE and discusses the properties of solutions that are essential to understand for their numerical solution. Many topics are streamlined here, too, with the time constraints of one term in mind. For example, in the treatment of boundary layers, only the derivation of the $O(\mathrm{Re}^{-\frac{1}{2}})$ estimate of the width of a laminar boundary layer is given. This estimate is important to understand for mesh generation and for estimating Re dependence of errors in a simulation. The boundary layer equations are omitted although they are so very close at hand after deriving this estimate.

Chapter 6 presents the essential theory of the steady NSE. The stability (meaning here that velocity is bounded by body force) of the velocity in the steady NSE is the connection between the mathematical architecture of the steady NSE and the physics of fluid motion. Uniqueness of the steady solution for small data is proved in Chapter 6 in a manner that introduces the steps in the convergence proof of Chapter 7 in a simplified setting.

The numerical analysis of the steady NSE is developed in Chapter 7 as a natural evolution of the stability bound for the velocity presented in Chapter 6. This convergence proof is essentially a simplification of the one in Girault and Raviart's wonderful 1976 monograph. The constants in the final result are not as sharp as those in the proof by Girault and Raviart because of these simplifications. For most students at this stage of their studies, this convergence proof will be the most complex one they will have struggled with. It is developed in steps in Chapter 7 and simplified to help students begin to see it as an

[4]One math student, who is now an accomplished applied analyst, told me that the part of the course that helped him the most was learning the summation convention!

elaboration of a few simple themes. In particular, it is presented as a variation on the proof of stability of the velocity and pressure of the continuous problem. Finally, the small data condition is interpreted for the time-dependent NSE.

Part III considers time-dependent fluid flow beginning with a summary of the Leray theory of the NSE in Chapter 8. It is my experience that proofs can be postponed, but any further simplification of the presentation of the theory results in much confusion that requires more time to correct later than is saved in the present. There are no useful shortcuts here: energy inequality, weak solution, strong solution, and uniqueness conditions are all needed, and all address essential physical issues relevant for computations. The Leray theory is built upon the physical foundation of the energy equality

$$\text{kinetic energy}(t) + \text{ total energy dissipated over } [0, t] \\ = \text{kinetic energy}(0) + \text{ total power input over } [0, t].$$

This is the most important and direct connection between fluid flow phenomena and the abstract Leray theory of the NSE. Chapter 8 proves stability of finite element methods for the time-dependent NSE by showing that the approximate solution satisfies the above energy equality. Convergence is studied in the energy norm. This is the most fundamental convergence analysis. It is a natural extension of the energy equality from the Leray theory of Chapter 8.

At this point, the topics could naturally end. One final topic is presented from among the enormous variety of fluid flow phenomena at the leading edge of CFD: turbulence. Some accepted physical theories of turbulence are easily accessible to students. These have not yet had impact in the numerical analysis community matching their physical importance and the insight they provide. Turbulence is also one of my own fascinations and research interests. I have restrained (with difficulty) the presentation of turbulence in Chapter 10 to homogeneous, isotropic turbulence and eddy viscosity models. Although not the leading edge, these are core ideas which still influence much current research. Indeed, much of the current research on numerical simulation of turbulent flows is aimed at attaining the good stability properties of (discretizations of) eddy viscosity models while avoiding their ad hoc nature, over damped effects on solutions and inaccurate predictions of many turbulent flows.

The field of fluid mechanics is wonderfully diverse. Fluids comprise three of the four states of matter, and the equations of fluid motion provide good models of many solids that flow as well (such as traffic and granular materials). There are also many materials that are important for industrial processing and manufacturing that sometimes behave like a solid and sometimes like a fluid! Laminar, isothermal, internal flows of a single, homogeneous, Newtonian liquid form only a fraction of fluid flows for which accurate predictions are needed. More complex flows add many layers of difficulties on top of those considered herein. Nevertheless, the understanding of incompressible, viscous flows is essential for progressing to more complex flows. There are also very many open questions and uncertainties in the numerical simulation of the case of laminar, isothermal, internal flows of a single, homogeneous, Newtonian liquid!

> *If nature were not beautiful, it would not be worth studying it. And life would not be worth living.*
>
> J. H. Poincaré, 1854–1912, quoted in G.W. Flake, *The Computational Beauty of Nature*, M.I.T. Press, 2000.

Acknowledgments

It is a pleasure to thank many people for their help in developing these notes. The reviewers of an early version provided detailed comments, far beyond normal expectations in a usual review. I appreciate their reviews, which were very helpful and definitely improved this book. My friends Vince Ervin, John Burkardt, and Noel Walkington gave me a lot of useful feedback (for which I am grateful) on various versions of this book and for some of the examples of FEM meshes (provided by Vince), the derivation of the energy equation in Example 3.3.6 (by Noel), and figures of finite element basis functions (by John). Other interesting and important figures were provided by Carolina Manica (the graph of energy dissipation rate versus kinetic energy), Leo Rebholz (some three-dimensional flow visualizations), Songul Kaya-Merdan (the driven cavity example), and Monika Neda (all the examples of flow around cylinders and the step flows simulations). I am grateful to Milan Jevtic, who helped by converting crude, hand-drawn schematics to professional-quality graphics. I appreciate all their help immensely. I thank my colleagues and friends at the University of Pittsburgh, Patrick Rabier, Paolo Galdi, Beatrice Riviere, Ivan Yotov, Chuck Hall, and Tom Porsching, for many vigorous and illuminating discussions on these topics. I owe a great debt to the graduate students upon whom the various versions of these notes were inflicted and a greater debt still to Max Gunzburger, who introduced me to the beautiful world of mathematical fluid dynamics years ago.

Part I

Mathematical Foundations

Chapter 1
Mathematical Preliminaries: Energy and Stress

Allez en avant, et la foi vous viendra.

(Go ahead and faith will come to you.)

d'Alembert, to a friend hesitant about infinitesimals, quoted in P. J. Davis and R. Hersh, *The Mathematical Experience*, Birkhäuser, Boston, 1981.

1.1 Finite Kinetic Energy: The Hilbert Space $L^2(\Omega)$

The laws of nature are drawn from experience, but to express them one needs a special language....

H. Poincaré, in *Analysis and Physics*, quoted in [21].

The function space $L^2(\Omega)$ consists of all velocities with *finite total kinetic energy* and is thus absolutely fundamental to mathematical fluid dynamics.

Example 1. A jet of water entering a large reservoir[5] at high speed is a common flow scenario; see Figure 1.1.[6]

A simplified realization of this flow involves a very large tank of water at rest which has a small opening on one side. A pipe leads to this opening and a jet of water can enter the tank through the opening. This is a classic flow problem with interesting behavior that is very easy to visualize by having the tank made of a clear material and putting dye in the jet. It is also very easy to control by increasing and decreasing the speed of the entering water.

The following is typically observed.[7] At low to moderate speeds, the jet enters and spreads out as it enters. Away from the jet, the water is very nearly at rest, and in the jet and

[5]The flow in Figure 1.1 does depict a jet entering a large resevoir, but it contains many features beyond that, including free surfaces, bubbly flow, turbulence, and quite complex geometries. The details of such flows are far from computationally predictable (and there are many open questions about how to best handle each of these complications). However, there are whitewater kayaking computer games in which simple (nonphysical) mathematical models produce animations of flows over waterfalls which pass the eyeball test. This is a good example of the difference in difficulty of description versus prediction.

[6]Photograph of the author by A. Layton.

[7]Note the word *typically*. Physical experiments on fluid motion are hard to perform accurately and even harder to do so that the same experiment with the same parameters gives the same flow.

Figure 1.1. *This flow is exciting but far beyond what is reliably computable!*

not too far from the entry point the water is very nearly at the entrance velocity. There is a fairly thin layer between the jet and the water where the two mix and in which the jet slows down and the water is dragged along by the jet. This mixing layer also spreads out as the jet flows along, and, farther from the jet entrance, both are fairly well mixed (and the jet is much slower). Still farther from the entrance, the water is well mixed and again at rest.

If the jet speeds up, it narrows and mixing and slowing takes longer. Above a certain speed, however, the mixing layer becomes quite complicated with swirling flow and the mixing properties of the flow increases again. Interestingly, if the high-speed swirling flow's pictures are averaged (for example, over some time interval), a very simple flow pattern reemerges that is quite like the flow entering the tank at low speeds!

This flow has many interesting features: mixing, layers, dissipation of energy, one fluid layer exerting a force on another and dragging it along, turbulent flow, and complicated mixing patterns which, upon averaging, resemble the flow at much slower speeds. Further, there are really only three ways to vary the experiment:[8]

1. fill the tank with a thicker and stickier (more viscous) fluid,
2. change the inlet velocity, and
3. make the inlet nozzle larger or smaller. ■

[8]A deeper analysis in Chapter 5 will show that there is really only one way to vary the experiment depending on one special combination of these parameters called the Reynolds number and defined by (diameter of orifice)∗(density of fluid)∗(inlet speed)/(viscosity of the fluid).

To make predictions about such flows requires a mathematical description of them. The flow of a jet into a tank (and most other flows), at the largest scales, has an input of kinetic energy; local differences in velocity exert a force upon adjacent parcels of fluids, altering the flow and dissipating energy. Thus, a mathematical description of this (and all other flows) requires at least two function spaces. The first is the space of all velocity fields with total kinetic energy finite. Since the complex patterns in fluid flows are (internally) created[9] not by the local velocity but by local velocity differences (i.e., velocity derivatives), the second space consists of all functions with velocity derivatives finite in a sense that is equivalent to the flow exerting a total finite force upon itself and dissipating in total a finite amount of energy. We shall introduce these two function spaces next!

Let Ω denote a domain (a bounded, open, connected set) with smooth enough boundary $\partial\Omega$ in $\mathbb{R}^2$ or $\mathbb{R}^3$ in which a fluid resides. The one space of functions that is essential for fluid mechanics is the Hilbert space $L^2(\Omega)$. Indeed, a recent and beautiful book by Doering and Gibbon [27] developed the essential mathematical theory of the Navier–Stokes equations (NSE) using only $L^2(\Omega)$ techniques. To understand its importance, suppose a fluid with constant density ρ_0 and velocity u is flowing in a domain Ω. In this setting, the total kinetic energy ($\frac{1}{2}$mass $\times$ velocity^2) is just

$$\text{kinetic energy} := \frac{1}{2}\rho_0 \int_\Omega |u|^2 \ dx.$$

The space $L^2(\Omega)$ is just the set of all velocity fields with finite kinetic energy.

Definition 1 ($L^2(\Omega)$ functions). *$L^2(\Omega)$ denotes the set of all functions $p : \Omega \to \mathbb{R}$ with*

$$\int_\Omega |p|^2 dx < \infty. \tag{1.1.1}$$

By the word "all" in Definition 1 we mean all Lebesgue measurable functions $p : \Omega \to \mathbb{R}$. There is another, equivalent, construction of $L^2(\Omega)$ that is useful. If we define a norm on the continuous functions $C^0(\Omega)$,

$$||p|| := \left[\int_\Omega |p(x)|^2 dx\right]^{1/2},$$

then $L^2(\Omega)$ can be defined as equivalence classes of Cauchy sequences:

$$then L^2(\Omega) = \text{ closure of } C^0(\Omega) \text{ in } ||\cdot||.$$

This definition of a space of functions by closure leaves open the interesting problem of characterizing the limit points. In $L^2(\Omega)$ these are precisely the Lebesgue measurable functions with $||p||$ finite. The norm $||\cdot||$ will herein *always* denote the $L^2(\Omega)$ norm.[10] The range of the function, i.e., d in $v : \Omega \to \mathbb{R}^d$, will usually be understood by the

[9] Often one hears the word *driven* used here (as in *fluid flow is driven by velocity differences*). More often and more correctly it is used to mean only the external forces supplying energy to the flow.

[10] This is the most common notation. However, another notation is seen in which $|\cdot|$ denotes the $L^2(\Omega)$ norm and $||\phi||$denotes the norm of $\nabla\phi$.

context in which $L^2(\Omega)$ is used. Sometimes it will be made explicit as in $L^2(\Omega)^d = L^2(\Omega) \times \cdots \times L^2(\Omega)$ (d times) for vector valued functions and $L^2(\Omega)^{d\times d}$ for $d \times d$ matrix valued functions.

Definition 2 (velocities with finite kinetic energy). $L^2(\Omega)^d = \{v = (v_1, \ldots, v_d) : \Omega \to \mathbb{R}^d$ *: each component* $v_j \in L^2(\Omega)$, $j = 1, \ldots d\}$ *and*

$$||v|| = ||v||_{L^2(\Omega)^d} := [||v_1||^2 + ||v_2||^2 + \cdots + ||v_d||^2]^{\frac{1}{2}}.$$

Lemma 1. $L^2(\Omega)^d = \{v = (v_1, \ldots, v_d) : \Omega \to \mathbb{R}^d : with\ ||v|| < \infty\}$.

Proof. Exercise. □

It is also necessary in mathematical fluid dynamics to consider a space of matrix-valued functions.

Definition 3 (tensors with finite kinetic energy). $L^2(\Omega)^{d\times d} := \{\mathbf{V} = (\mathbf{V}_{ij}), i, j = 1, \ldots, d : \mathbf{V}_{ij} \in L^2(\Omega)\}$, *i.e.,* $||\mathbf{V}|| < \infty$, *where*

$$||\mathbf{V}|| := \left[\sum_{i,j=1}^{d} ||\mathbf{V}_{ij}||^2\right]^{\frac{1}{2}}.$$

Again, often $L^2(\Omega)^{d\times d}$ will simply be written as $L^2(\Omega)$ when the range of the functions is clear. $L^2(\Omega)$ is a Hilbert space, i.e., a complete, normed, linear space whose norm is induced by an inner product $(\cdot, \cdot)$ given by

$$(p, q) := \int_\Omega p(x)q(x)dx \text{ for } p, q : \Omega \to \mathbb{R} \text{ and } p, q \in L^2(\Omega),$$

$$(u, v) := \int_\Omega \sum_{i=1}^{d} u_i v_i dx \text{ for } u, v \in L^2(\Omega)^d,$$

$$(\mathbf{S}, \mathbf{T}) := \int_\Omega \sum_{i,j=1}^{d} \mathbf{S}_{ij}(x)\mathbf{T}_{ij}(x)dx \text{ for } \mathbf{S}, \mathbf{T} \in L^2(\Omega)^{d\times d}$$

so that

$$||u|| = \sqrt{(u, u)} \ \forall u \in L^2(\Omega).$$

Definition 4. *Let* Ω *be a bounded domain. Then* $L_0^2(\Omega) \subset L^2(\Omega)$ *is defined by*

$$L_0^2(\Omega) = \left\{q : \Omega \to R : \int_\Omega q dx = 0, q \in L^2(\Omega)\right\}.$$

Lemma 2. $L_0^2(\Omega)$ *is a Hilbert space under the* $L^2(\Omega)$ *norm and inner product.*

Proof. Note that the function $r(x) = 1$ is in $L^2(\Omega)$ since Ω is bounded. The space

$$Y = span\ \{r(x)\}$$

is a (closed) one-dimensional subspace of $L^2(\Omega)$. Note now that by the definition of $L_0^2(\Omega)$,

$$L_0^2(\Omega) = \{q \in L^2(\Omega) : (q, 1) = 0\} = Y^{\perp}.$$

As an orthogonal complement, $Y^{\perp}$ is closed, and because Y is closed and properly inside $L^2(\Omega)$ so its complement is strictly inside $L^2(\Omega)$. Hence, $L_0^2(\Omega)$ is a closed subspace, strictly inside $L^2(\Omega)$. Thus, it is a Hilbert space under the inherited $L^2(\Omega)$ norm and inner product. □

The $L^2(\Omega)$ inner product satisfies the (very important) Cauchy–Schwarz and Young inequalities: for any $u, v \in L^2(\Omega)$

$$(u, v) \leq ||u||\ ||v|| \text{ and} \tag{1.1.2}$$

$$(u, v) \leq \frac{\epsilon}{2}\ ||u||^2 + \frac{1}{2\epsilon}\ ||v||^2 \text{ for any } \epsilon,\ 0 < \epsilon < \infty. \tag{1.1.3}$$

1.1.1 Other norms

For $1 \leq p \leq \infty$, the usual $L^p(\Omega)$ norm of a function $v : \Omega \to \mathbb{R}^d$ is defined by

$$||v||_{L^p} := \left[\int_\Omega |v|^p dx\right]^{1/p} \quad \text{if } 1 \leq p < \infty.$$

For example, the $L^4(\Omega)$ norm is

$$||v||_{L^4} := \left[\int_\Omega |v|^4 dx\right]^{1/4}.$$

The $L^\infty(\Omega)$ norm is the supremum over Ω excluding sets of measure zero. This is called the *essential supremum* and is written

$$||v||_{L^\infty} = \operatorname*{ess\ sup}_{x \in \Omega} |v(x)|.$$

The space $L^p(\Omega)$ can also be defined to be the closure of $C^0(\Omega)$ in the $L^p(\Omega)$ norm. It is a Hilbert space only for $p = 2$. The natural analogue of (1.1) and (1.2) are Hölder's and Young's inequalities: for any $\epsilon, 0 < \epsilon < \infty$, and $\frac{1}{p} + \frac{1}{q} = 1, 1 \leq p, q \leq \infty$,

$$(u, v) \leq ||u||_{L^p}||v||_{L^q} \text{ and } (u, v) \leq \frac{\epsilon}{p}||u||_{L^p}^p + \frac{\epsilon^{-(q/p)}}{q}\ ||v||_{L^q}^q. \tag{1.1.4}$$

1.2 Finite Stress: The Hilbert Space $X := H_0^1(\Omega)$

In mathematics, you don't understand things. You just get used to them.
J. von Neumann (1903–1957), quoted in G. Zukov, *The Dancing Wu Li Masters,* Rider and Co., 1979.

For the things of this world cannot be made known without a knowledge of mathematics.
R. Bacon, *Opus Majus,* Part 4, Distinctia Prima, Cap. 1, 1267.

The complex patterns in fluid flow are not created by large velocities but rather by large local changes in velocity, i.e., by first derivatives of u. The local changes in velocity are what cause one layer of fluid to exert a force or drag on the adjacent layer of fluid. The total force the fluid exerts upon itself in trying to get out of its own way must be finite. Thus, *if the velocity is to be physically relevant, its gradient must be in* $L^2(\Omega)^{d\times d}$. Thus, the second important function space consists of all vector functions with $\nabla v \in L^2(\Omega)^{d\times d}$.

Definition 5. *Let $d = dimension(\Omega) = 2$ or 3 . If $u = u_i, i = 1, \dots, d$, then ∇u is the $d \times d$ matrix of all possible first derivatives of u,*

$$(\nabla u)_{ij} = \frac{\partial u_j}{\partial x_i}, i, j = 1, \dots, d, \text{ and } ||\nabla u||^2 = \sum_{i,j=1}^{d} \left\| \frac{\partial u_j}{\partial x_i} \right\|^2 .$$

Further, $\nabla^s u$ is the symmetric part of ∇u,

$$(\nabla^s u)_{ij} = \frac{1}{2}\left(\frac{\partial u_j}{\partial x_i} + \frac{\partial u_i}{\partial x_j}\right), i, j = 1, \dots, d.$$

Let u be a $C^1(\Omega)$ function vanishing on $\partial\Omega$. Then,

$$||u||_X := [||u||^2 + ||\nabla u||^2]^{1/2}$$

is a norm which is induced by an inner product,

$$(u, v)_X := (u, v) + (\nabla u, \nabla v),$$

where $(\nabla u, \nabla v) = \sum_{i,j=1}^{d}(\frac{\partial u_i}{\partial x_j}, \frac{\partial v_i}{\partial x_j})$.

Definition 6. $X \equiv H_0^1(\Omega)$ *is the closure in* $||\cdot||_X$ *of*

$$\{v : \Omega \to \mathbb{R}^d : v \in C^1(\Omega) \text{ and } v = 0 \text{ on } \partial\Omega\}$$

Why do we require that velocities vanish on the boundary of the domain? This requirement is called the *no-slip condition*, and the requirement that $v = 0$ *on* $\partial\Omega$ is the appropriate expression of it when the boundary represents fixed, solid walls that are not moving. (This is often called an internal flow problem.) It is not at all obvious that this is the correct condition, and some great scientists have argued both yes and no. In a later chapter we shall explain the no-slip condition and its limitations. For now, we give a heuristic justification based on our experience with flow of air over our windshields.

Example 2 (the no-slip condition). After a rain, the windshield of your car is covered by many droplets of water. As you accelerate to highway speed, looking closely at them you will notice that the wind blasting over the windshield eventually makes the very large droplets smear. However, the smaller droplets, closer to the windshield, do not move at all: they stay stuck, and do not slip. This experiment is complicated by the surface tension of the drop and the drop's adhesion to the glass. However, another related experiment is also easy to do. Park under a tree in spring so that the glass will be covered by tree pollen. There is no pollen–glass adhesion and it is easily brushed off. However, in driving the car at normal speeds, the pollen is not blown off the windshield! This is strong evidence that the molecules of the fluid at a solid wall adhere to the wall and travel at the velocity of the wall. Thus the fluid velocity at a stationary wall must be zero! ∎

We now present (without proof) the characterization of element of X in terms of weak derivatives (introduced in Section 2.2).

Theorem 1. *With ∇u denoting the first order weak derivatives of u,*

$$X = \{u \in L^2(\Omega) : \nabla u \in L^2(\Omega) \text{ and } u|_{\partial\Omega} = 0 \text{ in } L^2(\partial\Omega)\}.$$

Remark 1. *We will often shorten the description of function spaces like X to, for example,*

$$X = \{u \in L^2(\Omega) : \nabla u \in L^2(\Omega) \text{ and } u = 0 \text{ on } \partial\Omega\}. \quad (1.2.1)$$

This is perfectly well defined if ∇u is interpreted (as we will do herein) as first order weak derivatives, and $u = 0$ on $\partial\Omega$ in the sense of the trace theorem. The expression "in the sense of the trace theorem" is often used and means in the sense of being zero everywhere on $\partial\Omega$ except on a set of $\partial\Omega$ measure zero.

The Poincaré–Friedrichs inequality will be used frequently herein. See Exercise 2 for the idea of its proof and Galdi [37] for a complete proof.

Theorem 2 (the Poincaré–Friedrichs' inequality). *There is a positive constant $C_{PF} = C_{PF}(\Omega)$ such that*

$$||u|| \le C_{PF}||\nabla u|| \; \forall u \in X. \quad (1.2.2)$$

To study less regular data, the $X^* = H^{-1}(\Omega)$ norm, $||\cdot||_{-1}$, is useful. Given a function $f \in L^2(\Omega)$ this X^* norm of f, $||f||_{-1}$ is defined by

$$||f||_{-1} = \sup_{v \in X} \frac{(f, v)}{||\nabla v||}.$$

Operationally, the X^* norm is the best norm for measuring the size of the body force f when we need to use a form of the Cauchy–Schwarz inequality resembling

$$(f, u) \le \{\text{some norm of } f\} \times ||\nabla u||.$$

Definition 7. *The function space $X^* = H^{-1}(\Omega)$ is the closure of $L^2(\Omega)$ in $||\cdot||_{-1}$.*

1.2.1 Weak derivatives and some useful inequalities

There is no science which did not develop from a knowledge of the phenomena; but in order to gain something from this knowledge, it is necessary to be a mathematician.

Daniel Bernoulli (1700–1782), in *Hydrodynamica*, Argentorati, 1738.

No human investigation can be called real science if it cannot be demonstrated mathematically.

L. da Vinci, in *Treatise on Painting*, 1651.

Since the function space X is defined by closure, the question of what the limit points are in X is an important one. To describe these limit points we must introduce the notion of a weak L^2 derivative. If $u \in C^1(\Omega)$, then for any function ϕ which is infinitely differentiable and vanishing in a strip near $\partial\Omega$, i.e., $\phi \in C_0^\infty(\Omega)$, the integration by parts formula

$$\int_\Omega \frac{\partial u}{\partial x_i}\phi\, dx = -\int_\Omega u \frac{\partial \phi}{\partial x_i} dx \text{ or } (u_{x_i}, \phi) = -(u, \phi_{x_i})$$

holds. The theory of weak or distributional derivatives has been highly developed and is motivated by this formula.

Definition 8. *Given $u \in L^2(\Omega)$, if there is a function $g(x) \in L^2(\Omega)$ satisfying*

$$(g, \phi) = -\left(u, \frac{\partial \phi}{\partial x_i}\right) \forall \phi \in C_0^\infty(\Omega),$$

then g is called the ith weak or distributional derivative of u. This is written $g = \frac{\partial u}{\partial x_i}$.

Since $\partial\Omega$ is a set of measure zero, functions in $L^2(\Omega)$ cannot in general have well-defined boundary values. On the other hand, there is an important theorem, called the trace theorem, stating that functions u with $u \in L^2(\Omega)$ and $\nabla u \in L^2(\Omega)$ do attain boundary values in a precise sense. We next present (without proof) one version of this important theorem.

Theorem 3 (the trace theorem). *Let $\partial\Omega$ be the graph of a Lipschitz continuous function. If $u \in L^2(\Omega)$ and $\nabla u \in L^2(\Omega)$, then $u|_{\partial\Omega}$ satisfies*

$$\begin{aligned} &u|_{\partial\Omega} \in L^2(\partial\Omega), \\ &||u||_{L^2(\partial\Omega)} \le C[||u||^2 + ||\nabla u||^2]^{1/2} = C||u||_X, \textit{ and} \\ &||u||_{L^2(\partial\Omega)} \le C||u||_X^{1/2}||u||^{1/2}. \end{aligned}$$

Thus, functions with weak derivatives in $L^2(\Omega)$ have well-defined boundary values. As a result, functions in X must vanish on $\partial\Omega$ (in the sense of being zero everywhere except on a set of $\partial\Omega$ measure zero). The trace theorem has many other important uses, which will become evident as the theory unfolds. For now, it is enough to know that it implies that functions in X have well defined boundary values and the norm of those boundary values is controlled by the norm in X on Ω.

The norm $||\nabla u||$ figures in many useful inequalities in analysis. The Ladyzhenskaya inequalities, below, are particularly interesting and useful. In them, the constants are absolute, i.e., they do not depend on the domain Ω.

Theorem 4 (the Ladyzhenskaya inequalities). *For any vector function $u : \mathbb{R}^d \to \mathbb{R}^d$ with compact support and with the indicated L^p norms finite,*

(i) $||u||_{L^4(R^2)} \leq 2^{1/4}||u||^{1/2}_{L^2(R^2)}||\nabla u||^{1/2}_{L^2(R^2)}$, $(d = 2)$,

(ii) $||u||_{L^4(R^3)} \leq \frac{4}{3\sqrt{3}}||u||^{1/4}||\nabla u||^{3/4}$, $(d = 3)$,

(iii) $||u||_{L^6(\mathbb{R}^3)} \leq \frac{2}{\sqrt{3}}||\nabla u||$, $(d = 3)$.

Proof. For the proof, see Ladyzhenskaya [62]. □

It is useful to consider functions f that are less regular than $f \in L^2(\Omega)$. This can occur, for example, if $f(x)$ has singularities which make $||f|| = \infty$, but $||f||_{L^p} < \infty$ for some p, $1 \leq p < 2$.

Proposition 1. *Let $\Omega \subset \mathbb{R}^2$ or $\mathbb{R}^3$. If $f \in L^2(\Omega)$, then $||f||_{-1} \leq C_{PF}||f|| < \infty$. Further, if $f \in L^p(\Omega)$ for $p \geq 6/5$, then $||f||_{-1} < \infty$.*

Because of the quadratic nonlinearity in the NSEs, useful upper bounds are needed for the product of three functions.

Lemma 3. *For any $u, v, w \in X$,*

$$\int_\Omega |u||v||w|dx \leq ||u||_{L^p}||v||_{L^q}||w||_{L^r}$$

for any p, q, r, $1 \leq p, q, r \leq \infty$, with $\frac{1}{p} + \frac{1}{q} + \frac{1}{r} = 1$, there is a $C = C(\Omega)$ such that

$$\left|\int_\Omega u \cdot \nabla v \cdot w dx\right| \leq C||\nabla u||||\nabla v||||\nabla w|| \text{ and}$$

$$\left|\int_\Omega u \cdot \nabla v \cdot w dx\right| \leq C||u||^{\frac{1}{2}}||\nabla u||^{\frac{1}{2}}||\nabla v||||\nabla w|| .$$

Proof. The first proof follows by repeatedly applying Hölder's inequality (exercise 7). The second follows from the first with $q = 2$, $p = 4$, $r = 4$, and the Ladyzhenskaya inequalities (Exercise 8). The third can be proved using similar tools. □

1.3 Some Snapshots in the History of the Equations of Fluid Motion

The history of fluid dynamics is replete with so many of the names of the great mathematicians that we can give only a few without turning the book into the field's history. Here are a few to spark your interest in the area.

Archimedes of Syracuse (287–212 BCE)

The notion of pressure is introduced in the first book on mathematical fluid mechanics, *Hydrostatics*. In fact, the most famous quote in science, "*Eureka!*," was in reference to his discovery of the principle of buoyancy: an immersed body is acted upon by a force equal to the weight of water it displaces.

After Archimedes, the basic ideas of dynamics took many years to develop and formalize. In particular, the idea of continuity of the fluid continuum was clearly stated by Leonardo da Vinci (1452–1519), who also performed detailed studies of waves, jets, and interacting eddies, and the concept of momentum in physics was introduced by Galileo (1564–1642).

I. Newton (1642–1727)

Newton introduced conservation of linear momentum (force $=$ mass $\times$ acceleration). He also studied and introduced the first concept of viscosity, *defectus lubricitatis*, for laminar flow as a special case of a linear stress–strain relation based on his experimental studies of fluid resistance.

D. Bernoulli (1700–1782)

Bernoulli's 1738 book, *Hydrodynamics*, is a major advance in mathematical fluid dynamics. In it, the first equations of fluid motion coupling the velocity and pressure are introduced, as is the kinetic theory of gasses, jet propulsion, and manometers. Bernoulli and Euler share credit for derivation of the famous Bernoulli equation

$$\frac{\rho v^2}{2} + p = \text{ constant},$$

which was the first explanation of the lift of airfoils.

L. Euler (1707–1783)

Euler is the father of fluid mechanics as a mathematical discipline. He derived the correct mathematical equations of inviscid flow now called the Euler equations, still used today, gave the complete and correct derivation of the Bernoulli equation, and even invented a hydraulic turbine. Lagrange wrote, "Euler did not contribute to fluid mechanics but created it."

After Euler's work, research in mathematical fluid mechanics accelerated rapidly. The next major milestone was the work of Navier and Stokes.

C. L. M. H. Navier (1785–1836) and G. Stokes (1819–1903)

In their work, the Navier–Stokes equations are derived and first written in their modern form. Navier first developed the equations of viscous flow. His final equations incorporated viscous effects correctly into the NSE but were justified based on an incorrect molecular model. Stokes gave the first clear and correct derivation of the viscous terms in the Navier–Stokes equations based on the Cauchy (1789–1857) stress principle. His derivation is similar to the one we use today. Others involved in the precise understanding of fluid resistance were Poisson (1781–1840) and Saint–Venant (1789-1886).

Navier and Stokes initially disagreed on the correct behavior of fluids on solid surfaces. This question was the heart of a scientific controversy with formidable natural philosophers on both sides of the issue. In retrospect, the resolution of the controversy was found already in 1879 by the great mathematical physicist James Clerk Maxwell.

J. C. Maxwell (1831–1879)

Maxwell is most famous for deriving the fundamental equations of electricity and magnetism, which are one of the great achievements of 19th-century science. However, he also had a keen interest in fluid mechanics. In 1879 in [70], Maxwell derived the continuum fluid flow equations by a limiting process beginning with Daniel Bernoulli's kinetic theory of gases. At the same time he studied correct conditions at a boundary and showed that, in a sense, both Navier (who argued for slip-with-friction at solid walls) and Stokes (who argued that the no-slip condition holds at walls) were correct. From the kinetic theory of gases he showed that the correct boundary conditions were

- no penetration: $u \cdot \widehat{n} = 0$; and
- slip velocity proportionate to tangential forces, or $u \cdot \widehat{\tau} + \beta \widehat{n} \cdot \nabla^s u \cdot \tau = 0$.

He showed that the friction coefficient scales like

$$\beta \sim \frac{\text{mean free path}}{\text{macro length scale}}.$$

In other words, in normal flows the correct condition was Stokes proposal of no slip: $u \cdot \widehat{\tau} = 0$. At the same time, Navier's boundary condition is also correct but the extra terms are significant only in regions of large stresses.

O. Reynolds (1842–1912)

Reynolds studied turbulence in real fluids experimentally. He showed the correct path to be prediction of turbulent flow averages rather than pointwise values. He also noted similarities between turbulent flows and flows with nonlinear viscosities. He solved the problem of how to upscale an experiment to a realistic flow geometry discovering the principle of a dynamic similarity with William Froude (1810–1879) and the key control parameter for the Navier–Stokes equations which we now call the Reynolds number.

H. Poincaré (1854–1912)

Poincaré is often described as the last great natural philosopher. He made singular contributions to many branches of mathematics, celestial mechanics, relativity theory, the philosophy of science, and fluid mechanics. Poincaré also introduced the method of sweeping—an early numerical method for solving partial differential equations—and forecasted the future arithmetization of analysis.

L. F. Richardson (1881–1953)

Richardson's book [81] gave insight into the physics of turbulent flows and computational complexity of turbulent flows simulation. In particular he introduced the concepts of eddy viscosity and the *energy cascade,* both scientifically and poetically. This concept provided strong motivation to the work of Kolmogorov which followed and gave a quantified description of the energy cascade in turbulence. Richardson was the first to apply finite difference methods to weather prediction.

J. Leray (1906–1998)

The founder of the mathematical theory of the Navier–Stokes equations was the great Jean Leray. In studying the Navier–Stokes equations, he introduced the idea of weak solution of partial differential equations which he called turbulent solutions, fixed point and degree theory through the Leray–Schauder fixed point theorem, connected topology to partial differential equations, and developed one of the major conjectures concerning turbulence. The basic estimates proved by Leray in his papers from 1934 are still essentially unimproved to this day. Leray has been described as the first modern analyst.

A. N. Kolmogorov (1903–1987)

A.N. Kolmogorov was an applied mathematician of depth, high originality, and breadth. In three short, clear, and simple papers, the theory of homogeneous, isotropic turbulence was introduced by Kolmogorov in 1941. The theory, now called the K41 theory, explains many universal features observed in turbulent flows. This work is a landmark of 20th-century science and bears the mark of genius.

J. von Neumann (1903–1957)

J. von Neumann is famous for his singular contributions in many areas of mathematics, including quantum theory, logic, ergodic theory, game theory, cellular automata, combustion, and computation. In his famous 1946 report, von Neumann summarized the status and future hopes of understanding turbulence. His assessment, which we shall cite elsewhere, is still valid today. His interest led him directly into computation as a method to extend our intuition into fluids. In a seminal paper in computational fluid dynamics, von Neumann and R. D. Richtmeyer developed numerical methods for shock problems. The methods are based, in part, on incorporating numerical realizations of nonlinear viscosities of the form now used today in large eddy simulation.

J. Smagorinsky (1924–2005)

Joseph Smagorinsky was a leader in the 1950s in using numerical methods and mathematical models to predict trends in weather and climate. He was the first to show that practical forecasting could really be done by solving the Navier–Stokes equations. Because of the limitations of the computers of the era, Smagorinsky was forced to struggle with the meaning of underresolved flow simulations. In 1963, Smagorinsky published a fundamental paper on numerical modeling of geophysical flow problems. Smagorinsky, on a different problem and for different reasons, independently rediscovered the von Neumann–Richtmeyer regularization of flow problems and used it in multidimensional problems. In geophysical large eddy simulation circles this is now known as the Smagorinsky model.

O. A. Ladyzhenskaya (1922–2004)

In the period 1964–1966, Ladyzhenskaya, inspired by the work of Jean Leray and possible connections between turbulence, large stresses, the linear stress strain relation, and breakdown of uniqueness of solutions to the three-dimensional Navier–Stokes equations independently added a correction term to the Navier–Stokes equations to account for possible nonlinear effects in a stress–strain relation. She developed complete, beautiful, and fully rigorous mathematical theory for the resulting system. In mathematical circles this model is often known as the Ladyzhenskaya model of fluid motion. Ladyzhenskaya wrote one of the fundamental books on the mathematical theory of the Navier–Stokes equations [62] as well as other important books on partial differential equations. Her work on mathematical fluid mechanics and nonlinear partial differential equations is of such prominence that she is widely considered to be one of the great analysts of the 20th century.

1.4 Remarks on Chapter 1

The mathematical analysis of the Navier–Stokes equations is a very deep area. This chapter introduces the basic function spaces used in its study. The natural next step to this chapter (beyond basic real analysis) is Chapter II of the wonderful book by Galdi [37]. Another reference that is a delight to study, currently interesting, and historically important is the 1969 book by Ladyzhenskaya [62]. The no-slip condition is only one of a large number of physically important boundary conditions in fluid flow problems. Its correctness was the subject of intense scientific debate, e.g., [22].

1.5 Exercises

Exercise 1. *It is known that for any two real numbers* a, b*,*

$$ab \leq \frac{a^p}{p} + \frac{b^q}{q}, \quad 1 \leq p, q \leq \infty, \quad \frac{1}{p} + \frac{1}{q} = 1.$$

Use this to prove Young's inequality from Hölder's inequality and (1.1.3) *from* (1.1.2).

Exercise 2. *Prove Theorem* 2 *for* $\Omega = (a, b)$ *an interval in* $\mathbb{R}^1$. ***Hint:*** *Since* $u(a) = 0 = u(b)$ *you can write* $u(x) = \int_a^x u'(s)ds$. *Thus,*

$$\int_a^b u^2(x)dx = \int_a^b \left[\int_a^x 1 \cdot u'(s)ds\right]^2 dx.$$

Now apply the Cauchy–Schwarz inequality.

Exercise 3. *Define* $L :=$ diameter(Ω). *Consider the Poincaré–Friedrichs inequality. By making a change of variables* $\hat{x}_j = x_j/L$ *map* Ω *into a domain* $\hat{\Omega}$ *of diameter* 1. *Do this on both sides of the inequality and find the dependence of* C_{PF} *on* $L = diam(\Omega)$.

Exercise 4. *Show that if* $u \in C^1(\Omega)$, *then the classical derivative* $\frac{\partial u}{\partial x_i}$ *is also the ith weak derivative of* u. *(This shows that writing* $g = \frac{\partial u}{\partial x_i}$ *is unambiguous.)*

Exercise 5. *Let* $\Omega = (-1, 1)$. *Show that the weak derivative of* $u = |x|$ *is* $u' = sign\ (x)$.

Exercise 6. *Prove Proposition* 1 *as follows. By Hölder's inequality*

$$(f, v) \leq ||f||_{L^p} ||v||_{L^q}.$$

Pick $p = 6/5$, $p^{-1}q = 6$ *(where* $\frac{1}{p} + \frac{1}{q} = 1$*) and apply the Ladyzhenskaya inequalities.*

Exercise 7. *Show that for any* p, q, r $, 1 \leq p, q, r \leq \infty$, *with*

$$\frac{1}{p} + \frac{1}{q} + \frac{1}{r} = 1,$$

it follows that

$$\int_\Omega fghdx \leq ||f||_{L^p} ||g||_{L^q} ||h||_{L^r}.$$

Hint: *Apply Hölder's inequality twice.*

Exercise 8. *Show that in three dimensions*

$$(u \cdot \nabla v, w) \leq C||u||^{\frac{1}{4}} ||\nabla u||^{\frac{3}{4}} ||\nabla v|| ||w||^{\frac{1}{4}} ||\nabla w||^{\frac{3}{4}}.$$

Hint: *Use Exercise* 7 *and the Ladyzhenskaya inequalities.*

Exercise 9. *Prove that in three and two dimensions if* $u \in X$, *then* $||u||_{L^4} \leq C||\nabla u||$.

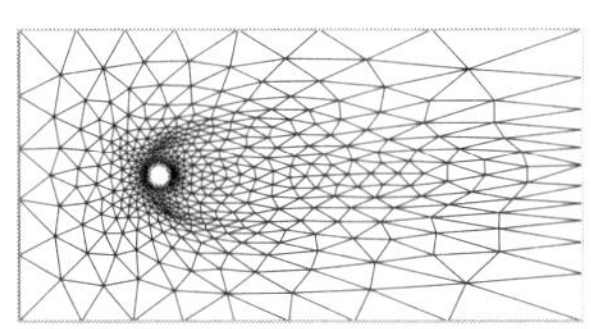

Chapter 2
Approximating Scalars

Whenever flexibility in the geometry is important—and the power of the computer is needed not only to solve *a system of equations, but also to* formulate *and* assemble *the discrete approximation in the first place—the finite element method has something to contribute.*
G. Strang and G. Fix, preface to [93].

2.1 Introduction to Finite Element Spaces

Cherchez la f.e.m.
G. Strang and G. Fix [93].

There are many problems in which all that is sought is a scalar distribution over space such as the temperature $u(x, y)$ at each point on a plate Ω. Another important example is flow of a liquid through a porous medium such as sand or soil. Such a flow is driven by the pressure pushing the liquid through the pores in the medium and predicting the flow depends on predicting the pressure. The flow's pressure is a scalar function which satisfies a partial differential equation much like the Poisson equation.

The pressure in an incompressible, above-ground flow also satisfies a Poisson problem whose data depend upon the flow velocity. Once the fluid velocity, u, is known, the pressure, $p(x, t)$, satisfies the *pressure Poisson equation*

$$\begin{aligned} -\Delta p(x,t) &= F(x,t) \text{ in the flow region } \Omega, \\ \nabla p(x,t) \cdot \widehat{n} &= g(x,t) \text{ on } \partial\Omega, \end{aligned}$$

where $\widehat{n}$ is the outward unit normal to $\partial\Omega$. The right-hand sides of this Poisson problem are determined by the fluid's velocity $u(x, t)$ and body force $f(x, t)$ by

$$\begin{aligned} F(x,t) &:= \sum_{i,j=1}^{d} \frac{\partial u_i}{\partial x_j}\frac{\partial u_j}{\partial x_i} - \operatorname{div} f(x,t)\,, \\ g(x,t) &:= (f(x,t) + \nu\Delta u)\cdot\widehat{n}\,. \end{aligned}$$

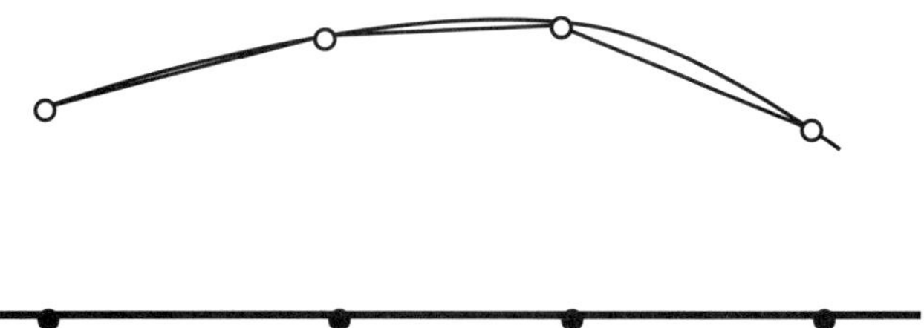

Figure 2.1. *A curve and its piecewise linear interpolant.*

(The derivation of the pressure Poisson equation must be postponed until the Navier–Stokes equations themselves are carefully derived.) Thus, one of the most basic operations in computational fluid dynamics (CFD) is to solve this (and many other) Poisson problems.

To begin this problem as simply as possible, consider the problem of representing as accurately as possible a known surface $z = u(x, y)$ defined on a polygonal, planar[11] domain Ω. We need to reduce an accurate approximate surface $u^h(x, y)$ representing $z = u(x, y)$ to one involving a finite number of degrees of freedom.

Example 3. In one dimension, the classic example of how a curve is represented by something more easily manipulated is by *linear interpolation*. Given an interval $\Omega = (a, b)$, points are selected on the interval (called *nodes* in finite element methods and knots in approximation theory):

$$a = x_0 < x_1 < x_2 < \cdots < x_N < x_{N+1} = b.$$

The piecewise linear interpolant, $u^h(x)$, is formed by connecting each $(x_j, u(x_j))$ with $(x_{j+1}, u(x_{j+1}))$ by a line segment, as depicted in Figure 2.1.

Associated with each node x_j, is a basis function $\phi_j(x)$, called a hat function (sketched in Figure 2.2).

With these basis functions the approximate curve $u^h(x)$ can be written explicitly as

$$u^h(x) = \sum_{j=0}^{N+1} u(x_j)\phi_j(x).$$

This representation and these basis functions are very important for finding an approximation to $u(x)$ when it is not known explicitly.

Filling in the complete details of this example comprises Exercise 11. ∎

In two dimensions, approximating a surface is done by introducing a triangulation $T^h(\Omega)$ and defining $u^h(x, y)$ on each triangle with a small number of degrees of freedom.

To begin constructing the approximate surface a triangulation $T^h(\Omega)$ is (somehow) constructed satisfying a few basic conditions:

[11]The case of a domain in three dimensions is, of course, very important, but it is harder to draw clear pictures in three dimensions illustrating the finite element method. Thus, in this chapter we focus on explaining the FEM for two-dimensional domains.

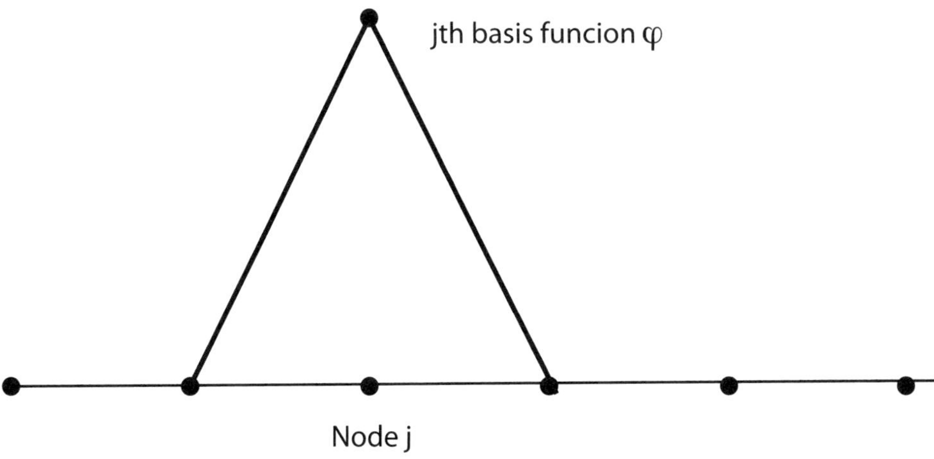

Figure 2.2. *A typical basis function.*

- *Conforming*: The triangles are all edge to edge, meaning a vertex of one triangle cannot lie on the edge of another.
- *Nondegeneracy:* The triangles are not close to straight line segments. This is measured in different ways. It is common to ask that the smallest angle in the triangulation be bounded away from either zero or the largest from 180 degrees.
- *The boundary is followed appropriately:* Generally this means that (i) the boundary of the computational domain is within the targeted error of the boundary of the real domain, and (ii) no triangle has all three vertices on a part of the boundary where Dirichlet boundary conditions are imposed.

If Ω is a polygon, then the general rule is that no triangle should have more than two vertices on $\partial\Omega$. (Often when nothing much interesting is happening in a corner or Neumann boundary conditions hold there, this general rule is overlooked, as in two corner triangles in Figure 2.3 and two in Figure 2.4. Of course, meshes are often reused for the same geometry and different boundary conditions so it is safest not to overlook this general rule.)

If the boundary contains a curved portion, then more care must be taken, as in the next example.

Often meshes are generated adaptively: a coarse mesh is selected, a solution on it computed, and some (automatic) calculation is performed to determine the regions of activity.[12] Next, the mesh is refined in those regions. With a good adaptive mesh generator, the final mesh can carry almost as much information as the solution. For example, Figures 2.5 and 2.6 are two adaptively generated meshes for flow around a cylinder that correspond to two different kinds of flows. The meshes reveal the structures of the solutions computed.

[12]Those regions are determined by a posteriori error estimators of the computed solution. A posteriori error estimation has become one of the central research areas in computational matematics and is having a big impact on how hard problems are actually solved.

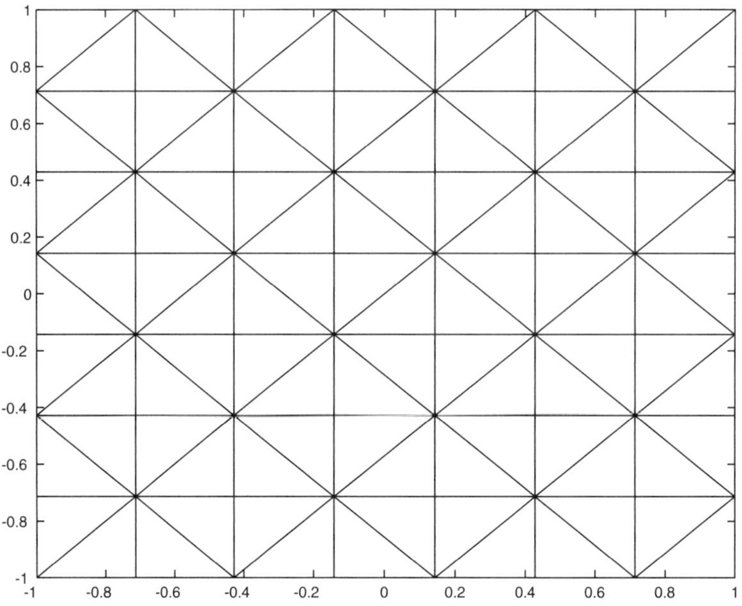

Figure 2.3. *Mesh with two "bad" triangles (upper left and lower right).*

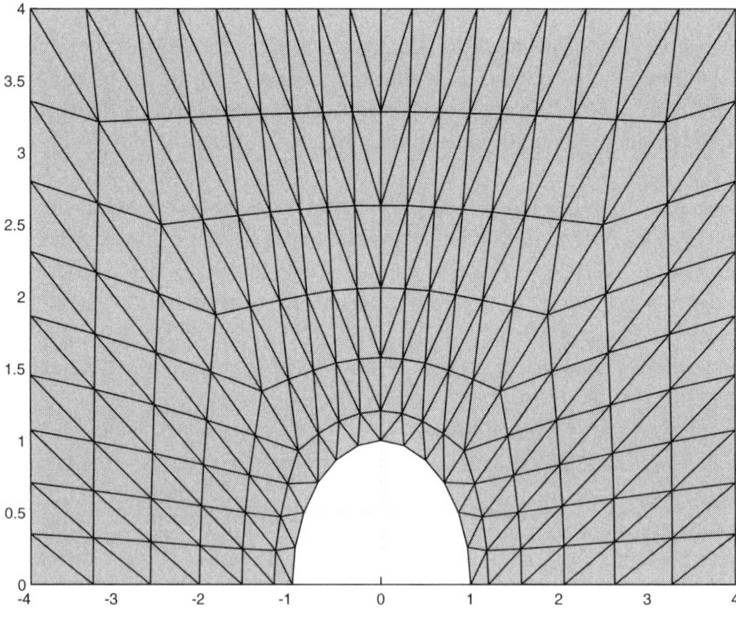

Figure 2.4. *Mesh following a curve.*

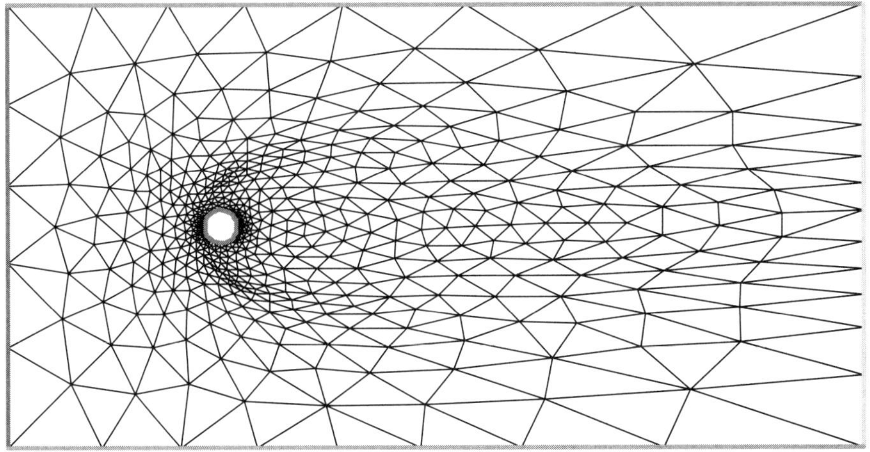

Figure 2.5. *Mesh for flow around cylinder at* Re = 40.

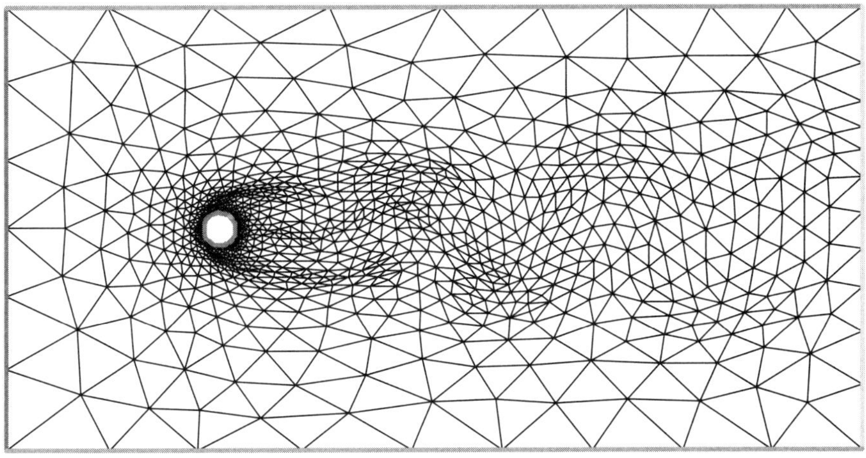

Figure 2.6. *Different mesh density for* Re = 200.

Once a mesh is generated, either by direct input or automatically, the approximations possible on that mesh must be selected. For the main example we are considering in this chapter, the approximate surface is globally continuous and a plane over each triangle and the *nodes* are just the vertices of $T^h(\Omega)$. The superscript h represents a triangle fineness measure:

$$h := \max_{all\ K \in T^h(\Omega)} \text{diameter}\,(K).$$

Define $X^h = X^h(T^h(\Omega))$, the set of approximate surfaces on $T^h(\Omega)$, by

$$X^h := \{\, u^h(x, y) \in C^0(\bar{\Omega}) : u^h \text{ is affine on each triangle } K,$$
$$u^h_{|_K}(x, y) = a_K + b_K x + c_K y, \text{ where } a_K, b_K, c_K \in \mathbb{R}\}.$$

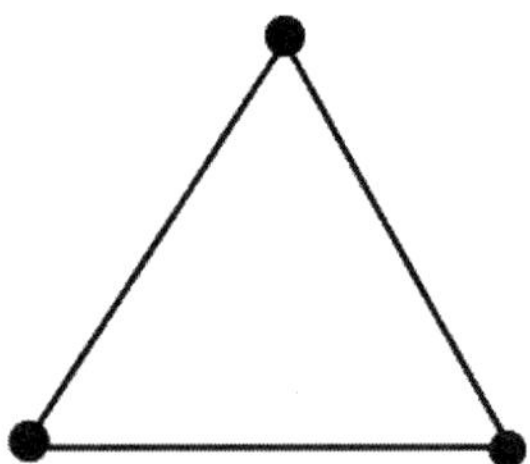

Figure 2.7. *The finite element space X^h.*

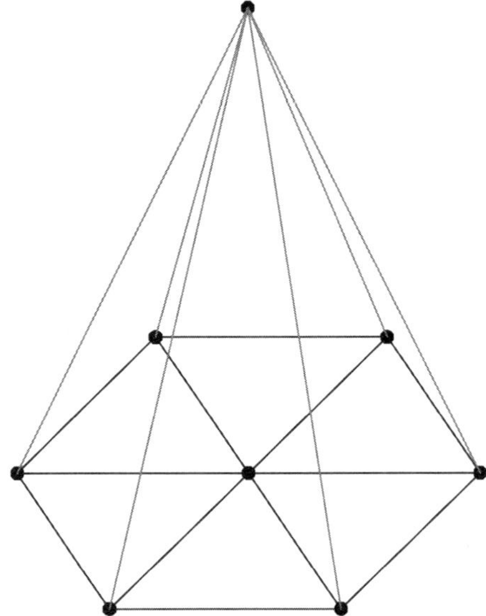

Figure 2.8. *A sketch of the basis function $\phi_j(x, y)$.*

We will shorten this to X^h and represent this space by Figure 2.7.

This representation is possible since such approximate surfaces are uniquely determined by their values at the vertices of $T^h(\Omega)$. Thus, we can associate each vertex of the triangle, marked in Figure 2.7 to indicate this association, with a degree of freedom in X^h, meaning a basis function for X^h.

The tent functions, depicted in Figures 2.8 and 2.9, form a basis for X^h. There is a 1:1 correspondence between these basis functions and the (vertex) nodes of $T^h(\Omega)$. This is indicated in Figure 2.9. Given a node N_j in $T^h(\Omega)$ the associated basis function $\phi_j(x, y)$ satisfies

$$\phi_j(x, y) = \begin{cases} 0 \text{ if } (x, y) \text{ does not lie on a triangle containing } N_j, \\ 1 \text{ if } (x, y) = N_j, \\ 0 \text{ at all other nodes } N_\ell,\ \ell \neq j, \\ \phi_j(x, y) \text{ is a plane (i.e., affine) inside each triangle.} \end{cases}$$

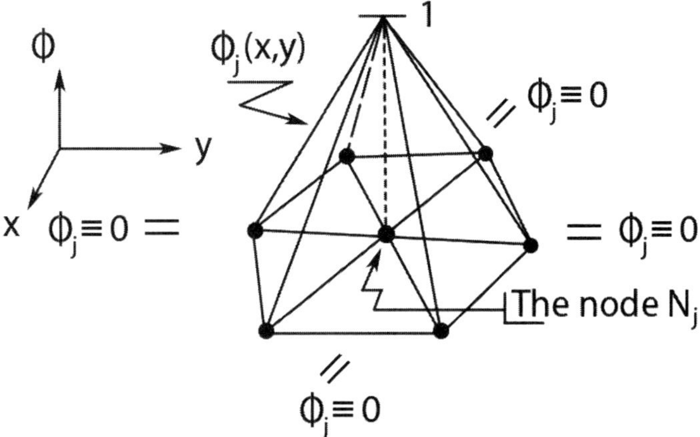

Figure 2.9. *The correspondence between basis functions and nodes.*

Any approximate surface $u^h(x, y)$ can then be written

$$u^h(x, y) = \sum_{\text{all nodes} N_j} c_j \phi_j(x, y),$$

where the (undetermined) coefficients c_j are just point values of u^h

$$c_j = u^h(N_j)$$

(provided these are known).

If the original surface $u(x, y)$ is known, we can use this property to approximate u by interpolation:

$$u(x, y) \cong (I_h u)(x, y) = u^h(x, y), \text{ where } u^h(x, y) = \sum_{all\ N_j} u(N_j)\phi_j(x, y).$$

This interpolant satisfies

$$u(N_j) = u^h(N_j).$$

It can be shown that u^h is close to u over all Ω if u is smooth enough (Theorem 2.1).

In some applications, it is important that u does not need to be smooth across inter-element edges, only inside each element.

Definition 9. *Let $u : \Omega \subset \mathbb{R}^d \to \mathbb{R}^1$. Then, $\nabla\nabla u$ is the $d \times d$ matrix*

$$\nabla\nabla u = \frac{\partial^2 u}{\partial x_i \partial x_j}, i, j = 1, \cdots, d. \tag{2.1.1}$$

The $L^2(\Omega)$ norm of $\nabla\nabla u$ is

$$||\nabla\nabla u|| := \left[\sum_{i,j=1}^{d} ||\frac{\partial^2 u}{\partial x_i \partial x_j}||^2\right]^{1/2}. \tag{2.1.2}$$

With this bit of notation we can summarize the interpolation error estimate for X^h.

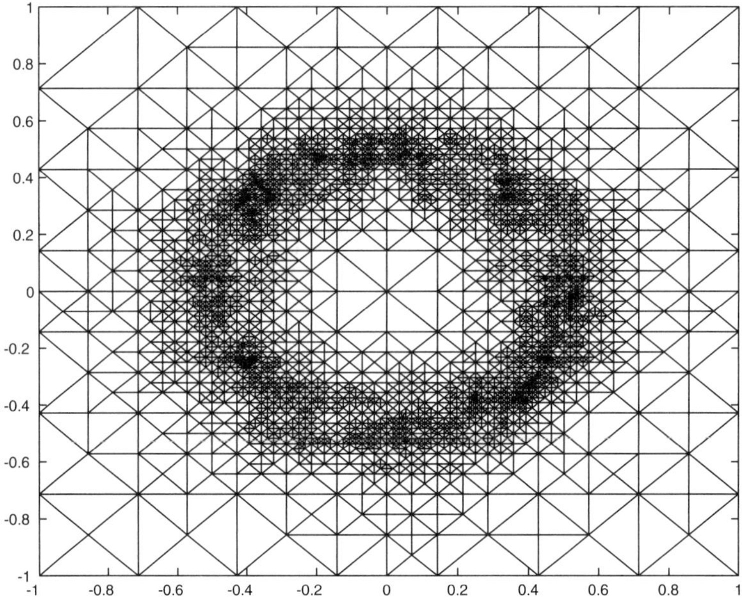

Figure 2.10. *A mesh resolving a circular transition region.*

Theorem 5 (interpolation error estimates). *Suppose that $u \in C^0(\overline{\Omega})$ and $\nabla\nabla u \in L^2(K)$ for every triangle $K \in T^h(\Omega)$. Then,*

$$||u - I^h(u)|| \le Ch^2 \left[\sum_{all\ K \in T^h(\Omega)} ||\nabla\nabla u||^2_{L^2(K)} \right]^{1/2},$$

$$||\nabla u - \nabla I^h(u)|| \le Ch \left[\sum_{all\ K \in T^h(\Omega)} ||\nabla\nabla u||^2_{L^2(K)} \right]^{1/2}.$$

Proof. The proof is proven in any good book on the finite element method. The book by Axelsson and Barker [7] (for example) gives a very clear direct proof. □

This estimates shows that for good accuracy, triangles should be small where $\nabla\nabla u$ is large. Generating such meshes means adapting the mesh size to estimates of the smoothness of the solution or of the local errors. It is nontrivial to do so and retain conformity of the meshes as well as ensure that no small angles are generated.[13] Fortunately, very good algorithms are known which can do this. For example, the mesh in Figure 2.10 (which contains two corner triangles that are bad for Dirichlet boundary conditions) was generated using an excellent algorithm of Maubach [69]. Every triangle in the mesh is a 45-45-90 triangle!

[13]It is nontrivial even when $\nabla\nabla u$ is known. It is harder of course when $\nabla\nabla u$ is unknown!

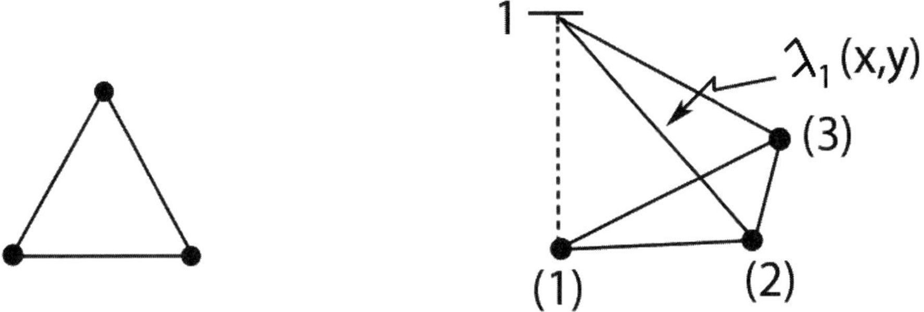

Figure 2.11. *One of the three linear basis functions.*

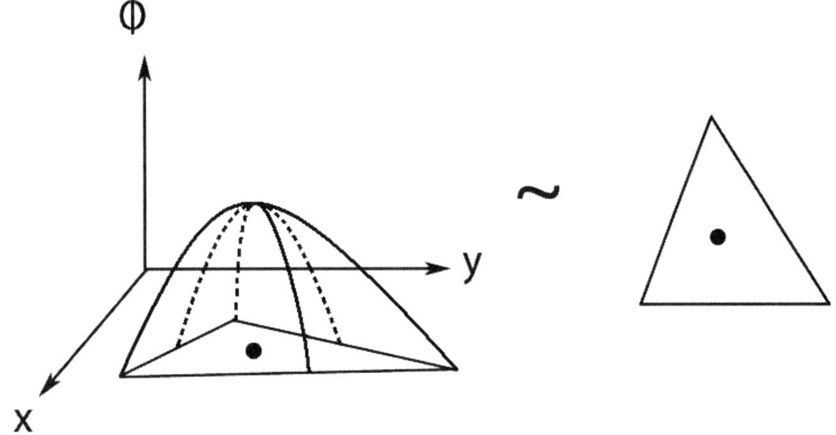

Figure 2.12. *The cubic bubble function.*

Naturally, if we *know* $u(x, y)$ we usually don't then *approximate* it. Still, this interpolation error estimate is important. It tells us the expected best we can do by approximation in X^h—even when we know $u(x, y)$ exactly at all the nodes.

The cubic bubble functions. Let X^h denote the space of C° piecewise linears on triangles. Each triangle $K \in T^h(\Omega)$ has three nodes N_1, N_2, and N_3. Let $\lambda_1(x, y)$, $\lambda_2(x, y)$, $\lambda_3(x, y)$ denote the associated basis functions, depicted in Figure 2.11.

With these three basis functions we can define a piecewise cubic that is nonzero on only one triangle—the *cubic bubble function*. Define

$$\phi_K(x, y) = \begin{cases} \lambda_1(x, y)\lambda_2(x, y)\lambda_3(x, y), (x, y) \in K, \\ 0, (x, y) \notin K. \end{cases}$$

It is easy to see that $\phi_K(x, y) \in C^\circ(\Omega)$, $\phi_K > 0$ on K and $\phi_K \equiv 0$ off K. The cubic function $\phi_K(x, y)$ is called the cubic bubble function and the associated node is the centroid of K, depicted in Figure 2.12.

2.2 An Elliptic Boundary Value Problem

If one looks at the different problems of the integral calculus which arise naturally when one wishes to go deep into the different parts of physics, it is impossible not to be struck by the analogies existing. Whether it be electrostatics or electrodynamics, the propagation of heat, optics, elasticity, or hydrodynamics, we are led always to differential equations of the same family.
H. Poincaré, in *American Journal of Physics* 12 (1890), 211.

Consider the problem of finding a surface $z = u(x, y)$ satisfying

$$-\nabla \cdot (k(x, y)\nabla u) = f(x, y) \text{ in } \Omega, \tag{2.2.1}$$

$$u = 0 \text{ on } \partial\Omega. \tag{2.2.2}$$

Here the functions f, k are known and the solution surfacc $u(x, y)$ is *not* known, so we cannot just interpolate u to find u^h. The coefficient $k(x, y)$ is positive and bounded

$$0 < k_0 \le k(x, y) \le k_1 < \infty,$$

$f \in L^2(\Omega)$, and Ω is a polygon in $\mathbb{R}^2$.

Example 4 (flow in porous media). This is already an interesting problem in fluid mechanics as the solution u can represent the *pressure* which pushes the flow through a porous media. We will postpone the details of the model for a bit and say only that in this application $k(x, y)$ describes the material the fluid flows through. For example, if sand contains a lump of clay, then k will be piecewise constant with an internal interface I on which k has a jump discontinuity, as depicted in Figure 2.13.

It is interesting to note in this example that since $k\nabla u$ must be differentiable across I, ∇u must be discontinuous across I to match the jump of $k(x, y)$ across I. ∎

Thus, the problem of approximating u has at least two parts:

1. $u(x, y)$ is not known. It must be extracted from the data and problem's physics.
2. $u(x, y)$ is not a classical solution of (2.2.1) since ∇u is discontinuous. Thus, we must work with a more general notion of solution.

Thus, the question is how to get a good approximation u^h of u in X^h without knowing u? The answer is: *the finite element method*! To present it, we first need to introduce a little notation. First, the variational formulation of the true solution is needed. To define it we will need the divergence theorem.

Lemma 4 (the divergence theorem). *If $\partial\Omega$ is smooth enough to be the graph of a Lipschitz function,*[14] *$\widehat{n}$ is its outward unit normal and $\overrightarrow{q}$, u, v are smooth enough, then*

[14]There is an intertesting technical question of how bizarre the boundary can be and still have the divergence theorem hold. A Lipschitz boundary is already quite general. Often, when this question is not central to the point under study, mathematicians skirt this issue by assuming that Ω is a "regular domain." A regular domain is any domain on which the divergence theorem is valid.

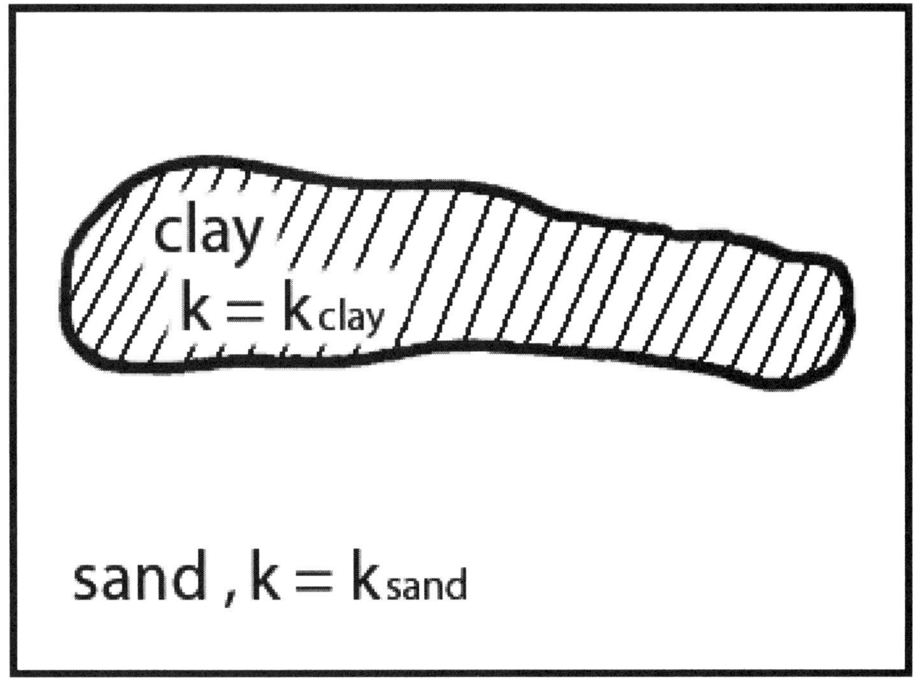

Figure 2.13. *An inhomogeneous medium.*

$$\int_{\Omega} \nabla \cdot \vec{q}\, dx = \int_{\partial\Omega} \vec{q} \cdot \widehat{n} ds, \tag{2.2.3}$$

$$\int_{\Omega} \frac{\partial u}{\partial x_i} dx = \int_{\partial\Omega} u n_i dx, \tag{2.2.4}$$

$$\int_{\Omega} \nabla \cdot (\nabla u v) = \int_{\partial\Omega} v \nabla u \cdot \widehat{n} ds = \int_{\Omega} (\Delta u) v + \nabla u \cdot \nabla v dx. \tag{2.2.5}$$

Proof. See almost any book on vector analysis. □

Let X denote, as usual, the function space:

$$X := \{v \in L^2(\Omega) : \nabla v \in L^2(\Omega) \text{ and } v = 0 \text{ on } \partial\Omega\}.$$

Recall that X is a Hilbert space under the inner product and norm:

$$(u, v)_X := \int_{\Omega} \nabla u \cdot \nabla v + uv dx \text{ and } ||v||_X = \sqrt{(v, v)_X}.$$

If we multiply (2.2.1) by $v \in X$ and integrate over Ω we readily see (after applying the divergence theorem) that the true solution of (2.2.1), (2.2.2), u, satisfies

$$\int_{\Omega} k \nabla u \cdot \nabla v dx = \int_{\Omega} f v \, dx \text{ for any } v \in X.$$

Definition 10. *The variational formulation of* (2.2.1) *is to find*

$$u \in X \text{ satisfying } (k\nabla u, \nabla v) = (f, v)\ \forall v \in X. \tag{2.2.6}$$

Exercise 10. *Find the variational formulation of the pressure Poisson equation. Be careful of the differences in the boundary conditions.*

The map $u, v \to (k\nabla u, \nabla v)$ (which is a real number) is linear in u for v fixed and linear in v for u fixed. This algebraic property is called bilinearity.

Definition 11 (bilinear forms). *The map* $a(\cdot,\cdot) : X \times X \to \mathbb{R}$ *is bilinear if for all* $u, v, w \in X$ *and all* $\alpha, \beta \in \mathbb{R}$,

$$a(\alpha u + \beta v, w) = \alpha a(u, w) + \beta a(v, w) \text{ and}$$
$$a(u, \alpha v + \beta w) = \alpha a(u, v) + \beta a(u, w).$$

Existence of a solution to (2.2.1) follows from its variational representation and the Lax–Milgram theorem, which we present next without proof. A proof can be found in any good book on finite element theory (and we have several very good ones listed at the end of this chapter) and in any good book on Hilbert space techniques.

Theorem 6 (the Lax–Milgram theorem). *Let* $a(\cdot,\cdot) : X \times X \to \mathbb{R}$ *be a bilinear form which satisfies*

$$(\text{continuity})\ a(u, v) \le a_1 ||u||_X ||v||_X\ \forall u, v \in X, \tag{2.2.7}$$

$$(\text{coercivity})\ a(u, u) \ge a_0 ||u||_X^2\ \forall u \in X \tag{2.2.8}$$

for some positive constants a_0, a_1. *Let* $F : X \to \mathbb{R}$ *be a linear functional which is*

$$(\text{continuous})\ F(v) \le C||v||_X\ \forall v \in X. \tag{2.2.9}$$

Then, there exists a unique $u \in X$ *satisfying*

$$a(u, v) = F(v)\ \forall v \in X. \tag{2.2.10}$$

Furthermore, $||u||_X \le C/a_o$.

Our immediate goal is to apply the Lax–Milgram theorem to the above formulation (2.2.6) of the elliptic boundary value problem (2.2.1). To do this requires verifying that

$$a(u, v) := (k\nabla u, \nabla v)$$

is continuous and coercive. Continuity is (we will see) easy, but coercivity requires an important mathematical result—the Poincaré–Friedrichs inequality, which we recall from Chapter 1.

Theorem 7 (Poincaré–Friedrichs inequality). *Let* $\Gamma_0 \subset \partial\Omega$ *have positive measure. Let*

$$X := \{v \in L^2(\Omega) : \nabla v \in L^2(\Omega) \text{ and } v = 0 \text{ on } \Gamma_0\}. \tag{2.2.11}$$

Then, there is a positive constant $C = C_{PF} > 0$ such that

$$||v|| \leq C_{PF}||\nabla v|| \; \forall v \in X. \tag{2.2.12}$$

Thus, $||\nabla v||$ and $||v||_X$ are equivalent norms on X.

Proof. We will prove only a special case which captures the *idea* of the proof without the notational intricacies. We will prove (2.2.12) for C^1 functions v which vanish on the entire boundary, $\Gamma_0 = \partial\Omega$.

If $\Omega = (a, b)$ is in $1 - d$, then we can write

$$v(x) = \int_a^x v'(t)\, dt, \text{ thus } v(x)^2 = \left(\int_a^x v'(t)dt\right)^2.$$

Thus, the Cauchy–Schwarz inequality gives

$$v(x)^2 = \left(\int_a^x 1 \cdot v'(t)dt\right)^2 \leq \left(\left(\int_a^x 1^2 dt\right)^{1/2} \left(\int_a^x v'(t)^2 dt\right)^{1/2}\right)^2 \leq (b-a)||v'||^2.$$

Integrating this inequality gives

$$||v||^2 = \int_a^b v(x)^2 dx \leq \int_a^b (b-a)||v'||^2 dx = (b-a)^2||v'||^2,$$

which is (2.2.12) with $C = \sqrt{b-a}$. If Ω is a rectangle in $2 - d$, $\Omega = (a, b) \times (c, d)$ with $\Gamma_0 = \partial\Omega$, then we can still write

$$v(x, y)^2 = \left(\int_a^x \frac{\partial v}{\partial x}(t, y)\, dt\right)^2 \leq \cdots \leq (b-a)\left\|\frac{\partial v}{\partial x}\right\|^2.$$

Repeating the above $1 - d$ argument proves the result for rectangles. For more general domains, a similar argument is used, writing $v(x)$ as a line integral from a point $a \in \Gamma$, to the point $x \in \Omega$. □

Using (2.2.12), it is easy to verify the conditions of the Lax–Milgram theorem and thus prove existence of a variational solution to the elliptic boundary value problem (2.2.1) or, equivalently, (2.2.6).

Proposition 2 (continuity and coercivity). *Let $0 < k_{\min} \leq k(x, y) \leq k_{\max} < \infty$, $a(u, v) = \int_\Omega k\nabla u \cdot \nabla v dx$ is continuous and coercive on X:*

$$a(u, u) = \int_\Omega k|\nabla u|^2 dx \geq Ck_{\min}||u||_X^2, \tag{2.2.13}$$

$$a(u, v) \leq k_{\max}||\nabla u||\; ||\nabla v|| \leq Ck_{\max}||u||_X||v||_X \forall v \in X. \tag{2.2.14}$$

Proof. Exercise 14. □

One consequence is that our elliptic problem is well-posed.

Corollary 1 (existence and uniqueness of a solution). *Let* $0 < k_{\min} \le k(x, y) \le k_{\max} < \infty$. *For any* $f \in L^2(\Omega)$ *the elliptic boundary value problem has a unique solution* $u \in X$ *satisfying* (2.6). *Further, there is a constant* $C = C(\Omega, k_{\min})$ *such that*

$$||\nabla u|| \le C||f||. \tag{2.2.15}$$

Actually, the Lax–Milgram theorem shows that equation (2.2.6) can be solved for a more general class of data f than $f \in L^2(\Omega)$. Specifically, it implies (2.6) is uniquely solvable for any $f \in H^{-1}(\Omega)$, a wider set of right-hand sides than $L^2(\Omega)$. To define $H^{-1}(\Omega)$, note that for $f \in L^2(\Omega)$ the dual norm (defined below) is finite:

$$||f||_{-1} := \sup_{v \in X} \frac{(f, v)}{||\nabla v||}.$$

Thus, $H^{-1}(\Omega) = X^*$, the set of all bounded linear functionals on X, can be defined by closure and will satisfy $X^* = H^{-1}(\Omega) \supset L^2(\Omega)$.

Definition 12 (the dual space of X). $X^* = H^{-1}(\Omega)$ *is the closure of* $L^2(\Omega)$ *in* $|| \cdot ||_{-1}$.

2.3 The Galerkin–Finite Element Method

Because the mathematical foundations are sound, it is possible to understand why the method works.
G. Strang and G. Fix, preface to [93].

The finite element method (FEM) is a numerical method for solving complex problems in engineering, science, and technology. It has proved to have great advantages when geometries become complex, models more complicated, and the behavior of the solution more uncertain. On the other hand, for simple problems in simple domains like squares, it is also more work to implement than, say, finite difference methods. The robustness and reliability of the FEM is a direct result of its sound mathematical foundation that connects (via its mathematical architecture) the fundamental physical principles of the problem to the discrete equations the FEM produces for the approximation of the problem.

The FEM is a combination of the Galerkin approximation and the choice of the right finite dimensional space X^h. We have seen that when starting with a partial differential equation, we can derive from it an equivalent variational formulation.On the other hand, often the partial differential equation itself arises from the physical problem being considered by some physical principle, such as virtual work, minimization of some physical energy, etc., that is precisely the variational formulation of the PDE model. One reason for the robustness of the FEM is that it begins with a variational formulation instead of a pointwise differential equation. In general, given a bounded linear function $F \in X^*$, the problem is to find $u \in X$ satisfying

$$a(u, v) = F(v) \ \forall\, v \in X. \tag{2.3.1}$$

The bilinear form $a(u, v)$ is assumed (as in the motivating application (2.2.1)) to be continuous and coercive. The Galerkin method begins by selecting a finite dimensional subspace $X^h \subset X$. The Galerkin approximation $u^h \in X^h$ is the solution of the equations: find $u^h \in X^h$ satisfying

$$a(u^h, v^h) = F(v^h) \ \forall\, v^h \in X^h. \tag{2.3.2}$$

At this level of abstraction, there is a complete convergence theory for the Galerkin method given in Céa's lemma.

Theorem 8 (Céa's lemma). *Suppose $a(\cdot,\cdot): X \times X \to R$ is a continuous and coercive bilinear form and $F : X \to R$ is a bounded linear functional. Let $X^h \subset X$ be a finite dimensional subspace of X. Then, there is a unique $u^h \in X^h$ satisfying*

$$a(u^h, v^h) = F(v^h) \ \forall v^h \in X^h. \tag{2.3.3}$$

The error $u - u^h$ satisfies[15]

$$||u - u^h||_X \le \left(1 + \frac{a_1}{a_0}\right) \inf_{v^h \in X^h} ||u - v^h||_X. \tag{2.3.4}$$

Proof. Since X^h is finite dimensional, existence and uniqueness of u^h are implied by the associated linear system having a zero null space, which is implied by coercivity. For the error estimate, let $e = u - u^h$ and note that

$$\text{Galerkin Orthogonality: } a(e, v^h) = 0 \ \forall\, v^h \in X^h$$

holds by subtracting (2.3.1) and (2.3.2). Let $v^h \in X^h$ be given and decompose

$$e = u - u^h = (u - v^h) - (u^h - v^h) = \eta - \phi^h,$$

where $\eta = u - u^h$, $\phi^h = u^h - v^h$. Note that $\phi^h \in X^h$. We think of ϕ^h as the piece of the error in X^h and η as the piece outside of X^h.

Now $a(\eta - \phi^h, v^h) = 0 \ \forall\, v^h \in X^h$. Thus

$$a(\phi^h, v^h) = a(\eta, v^h),$$

setting $v^h = \phi^h$, and using coercivity and continuity gives

$$a_0||\phi^h||_X^2 \le a(\phi^h, \phi^h) = a(\eta, \phi^h) \le a_1||\eta||_X||\phi^h||_X. \tag{2.3.5}$$

Thus, canceling the common factor from both sides,

$$||\phi^h||_X \le \frac{a_1}{a_0}||\eta||_X.$$

The triangle inequality now implies

$$||u - u^h||_X \le ||\eta||_X + ||\phi^h||_X \le \left(1 + \frac{a_1}{a_0}\right)||\eta||_X = \left(1 + \frac{a_1}{a_0}\right)||u - v^h||_X.$$

Taking the infimum over $v^h \in X^h$ yields the result. □

[15]The "1+" in this estimate can be eliminated by a slightly different proof, Exercise 18. The proof given seems to be the one whose structure is easiest to generalize. Nevertheless, different extensions beyond the assumptions of Céa's lemma can require different strategies to get around their difficulties. The exercises include three different proofs, all of which have proved useful.

Let us now briefly consider the linear system coming out of the Galerkin method. Pick a basis $\{\phi_j : j = 1, \ldots, N\}$ for X^h and expand

$$u^h = \sum_{j=1}^{N} c_j \phi_j, \; (c_j \in \mathbb{R}),$$

thus finding u^h is equivalent to finding the undetermined coefficients c_j. Inserting this expansion into (2.3.2) and setting $v = \phi_k$ gives the equations for the unknown coefficients c_j,

$$\sum_{j=1}^{N} c_j \, a(\phi_j, \phi_k) = (f, \phi_k), \; k = 1, \ldots, N.$$

This is an $N \times N$ linear system for the coefficients c_j:

$$A_{N \times N} \vec{c} = \vec{f} \; A_{ij} = a(\phi_i, \phi_j). \tag{2.3.6}$$

Calculating u^h is equivalent to solving this nonsingular $N \times N$ linear system. This matrix A is known as the stiffness matrix.

The second ingredient in the FEM is the precise and clever choice of the space X^h (as in the first section). This must be chosen for *accuracy* (the error $\inf_{X^h} ||u - v^h||_X$ is small) and *efficiency* (the linear system (2.3.6) is sparse and structured). Although we will not discuss efficiency herein, the accuracy of the FEM is a direct corollary of Céa's lemma and Theorem 5.

Proposition 3 (convergence). *Let X^h be the space of C^0 piecewise linears on a triangulation $T^h(\Omega)$. Then, the finite element approximation u^h of the solution u of* (2.2.1) *satisfies:*

$$\begin{aligned}
&[||\nabla(u - u^h)||^2 + ||u - u^h||^2]^{1/2} \\
&\leq C(1 + k_{\max}/k_{\min}) \inf_{v^h \in X^h} [||\nabla(u - v^h)||^2 + ||u - v^h||^2]^{1/2} \\
&\leq C\left(1 + \frac{k_{\max}}{k_{\min}}\right) h||\nabla\nabla u||.
\end{aligned}$$

Remark. The same result holds with $||\nabla\nabla u||$ replaced by

$$\left(\sum_{all \; K \in T^h(\Omega)} ||\nabla\nabla u||^2_{L^2(K)} \right)^{1/2}.$$

This estimate also suggests that the mesh should be small where $\nabla\nabla u$ is large. Unfortunately, unlike in interpolation, $\nabla\nabla u$ is unknown! Other types of error estimates, called a posteriori error estimates, have been developed to automate decisions of where mesh refinement should take place. As one example, the mesh that arises from one such process for a solution that looks like a step function skew to the domain is pictured in Figure 2.14. This mesh, like the previous example, is generated by Maubach's algorithm [69]. Every triangle in the mesh is a 45-45-90 triangle!

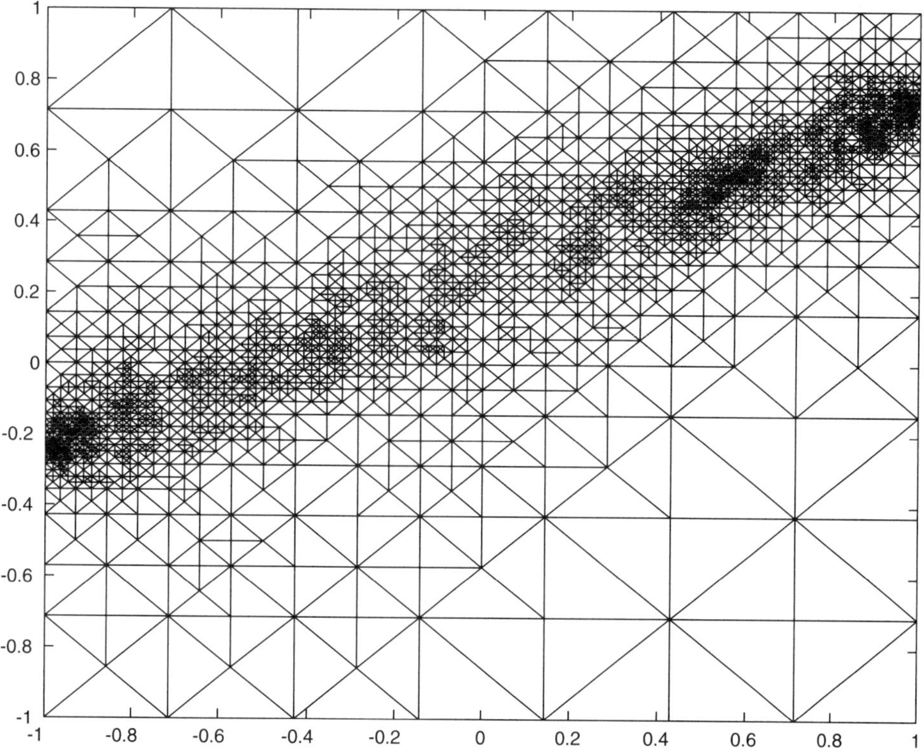

Figure 2.14. *A typical adaptive FEM mesh.*

2.4 Remarks on Chapter 2

There has been time only for the briefest introduction to the FEM for scalar problems. The presentation in Chapter 2 was done to stress only the aspects of the FEM that are directly used in later chapters—not the parts most important either computationally or mathematically. Almost every applied mathematics program and engineering school offers good courses on FEMs. The literature on FEMs is enormous. Among the many good books on the topic, three come to mind as accessible to students, mathematically interesting, and complementary in focus: those of Axelsson and Barker [7], Johnson [61], and Strang and Fix [93].

The analysis in this chapter has focused on stability of the approximate solution, i.e., on finding conditions under which the approximate solution is bounded in a physically relevant norm by the problem data. Convergence and estimates of the error follow closely from stability. These estimates are known as á priori error estimates. Such error estimates show asymptotic convergence as $h \to 0$ and, when carefully done, give the asymptotic rate of convergence. On the other hand, since the right-hand side involves the unknown true solution, á priori error estimates cannot be used to give a numerical bound on the error or to decide where the mesh should be coarse and where it should be fine. The complementary type of error estimate, a posteriori estimates, give a computable upper bound to the error involving (ideally) only the known approximate solution on the right-hand side. When the computable upper estimate is a sum over triangles of quantities which are computable on each triangle, a posteriori error estimates can be used to decide where to refine the FEM mesh. (How to do it, meaning mesh generation and refinement, has grown into a research area of its own; most of the meshes we displayed were generated adaptively using an algorithm of Maubach [69].) A posteriori error estimation is an important contribution that numerical analysis has made to practical computing. For more information on a posteriori estimation, the books of Ainsworth and Oden [1] and Verfürth [100] are good paces to start reading.

Generalizations of the Lax–Milgram theorem in which coercivity is replaced by an inf-sup condition are very important, and the central one is due to Babuska. It replaces coercivity with an inf-sup condition. This generalization can be used to develop a proof of the existence of the pressure in incompressible flows; see Chapter 4.

Theorem 9 (Babuska's extension of the Lax–Milgram theorem). *Let H_1, H_2 be real Hilbert spaces, $a(\cdot,\cdot) : H_1 \times H_2 \to \mathbb{R}$ a bilinear form, and $F : H_2 \to R$ a bounded linear functional. Suppose there are positive constants C_1, C_2 such that*

$$|a(u,v)| \leq C_1 ||u||_{H_1} ||v||_{H_2} \;\; \forall u \in H_1, v \in H_2, \tag{2.4.1}$$

$$inf_{u \in H_1} sup_{v \in H_2} \frac{|a(u,v)|}{||u||_{H_1} ||v||_{H_2}} \geq C_2 > 0, \tag{2.4.2}$$

$$sup_{u \in H_1} |a(u,v)| > 0 \;\; \forall 0 \neq v \in H_2. \tag{2.4.3}$$

Then, there is a unique $u_0 \in H_1$ satisfying $a(u_0, v) = F(v) \;\; \forall v \in H_2$ and, furthermore, $||u_0|| \leq C_2^{-1} ||F||_{H_2}$.

The error analysis in the energy norm (i.e., in X) is direct, complete, and mathematically pleasing. Nevertheless, other norms are useful and important. Estimates in other norms are obtained by a tool called duality, which should be the first step in readings on finite element theory beyond this chapter, e.g., in [7], [61]. Another very important current direction in finite element analysis is a posteriori error estimation and its use for adapting meshes.

2.5 Exercises

Exercise 11 (piecewise linear finite elements in one dimension). *Develop the theory of continuous, piecewise linear finite elements in one dimension exactly paralleling the theory of this section in two dimensions. A domain in one dimension is an interval (a, b), and a mesh in one dimension is a collection of points (the nodes):*

$$a = x_0 < x_1 < x_2 < \cdots < x_N < x_{N+1} = b.$$

The C^0 piecewise linears are

$$\{v^h(x) \in C^0[a,b] : v^h(x) \text{ is affine on each } [x_j, x_{j+1}]\}.$$

(a) *Sketch the interpolant of a typical function* $v(x)$ *(it is like connecting the dots).*

(b) *Sketch the basis functions. There is one basis function for each node. (These resemble witches hats and are usually called hat functions.)*

(c) *Show that the hat functions are linearly independent. (**Hint:** Write down a dependency relation then evaluate it at a node.)*

(d) *Show that the hat functions span the finite element space by a dimension argument. (**Hint:** Count the degrees of freedom in the space and the number of hat functions.)*

Exercise 12 (piecewise quadratics in one dimension). *Repeat the previous exercise for the space of* C° *piecewise quadratic functions in one dimension. These have three nodes per element (a subinterval) the two vertices and the midpoint. Consulting books on finite element methods, write up a short description of this space. Include* (i) *a sketch of the basis functions and* (ii) *a statement of the known bounds on the error in interpolation in* X^h.

Exercise 13 (piecewise quadratics in two dimensions). *The space of* C° *piecewise quadratic functions defined on a triangulation of the domain is represented by a triangle with six nodes: the three vertices and the three midedges. Consulting books on finite element methods, write up a short description of this space. Include* (i) *a sketch of the basis functions and* (ii) *a statement of the known bounds on the error in interpolation in* X^h.

Exercise 14. *Prove Proposition* 2.

Exercise 15. *Prove that* $||\nabla u||$ *and* $||u||_X$ *are equivalent norms on* X *if* (2.2.12) *holds.*

Exercise 16. *Consider the Lax–Milgram theorem. Let the correspondence between right-hand-side* f *and solution* u *be denoted by* $T(f) = u$. *Show that* T *is linear.*

Exercise 17. *Consider the variational solution of* (2.2.1) *with* $f \in X^*$. *Define* $T(f) = u$. *Show that* $T : X^* \to X$ *is, in addition to being linear, a bounded operator.*

Exercise 18 (another proof of Cea's lemma). *In step* (2.3.5), *in the proof of Céa's lemma instead of canceling the* $||\phi^h||_X$ *expand the right-hand side using*

$$||\eta||_X||\phi^h||_X \leq \frac{\epsilon}{2}||\phi^h||_X^2 + \frac{1}{2\epsilon}||\eta||_X^2, \; 0 < \epsilon < 1,$$

and complete the proof from that point.

Exercise 19 (a third proof of Céa's lemma with better error constant). *Complete the following (third) proof of Céa's lemma:*

$$a_0||u - u^h||_X^2 \leq a(u - u^h, u - u^h) = a(u - u^h, u - v^h) \text{ for any } v^h \in X^h.$$

Explain why the above is true, and complete the proof using continuity. This proof leads to a better error constant.

Exercise 20. *Prove that $u^h \in X^h$ exists by applying the Lax–Milgram theorem in X^h.*

Exercise 21. *Show that $A_{N\times N}$ is positive definite and hence nonsingular.*

Exercise 22 (the Galerkin projection). *Consider the linear operator $P^h : u \to u^h$ where u^h satisfies*

$$a(u - u^h, v^h) = 0 \;\forall\; v^h \in X^h.$$

Show $P^h : X \to X^h$ satisfies

$$||P^h||_{\mathcal{L}(\mathcal{X}\to\mathcal{X}^{\langle})} \le C \text{ and } P^h\ P^h u = P^h u.$$

In other words, P^h is a projection operator with uniformly bounded norm. (A fourth proof of Céa's lemma can be based on this result.)

Exercise 23. *Prove Proposition* 3.

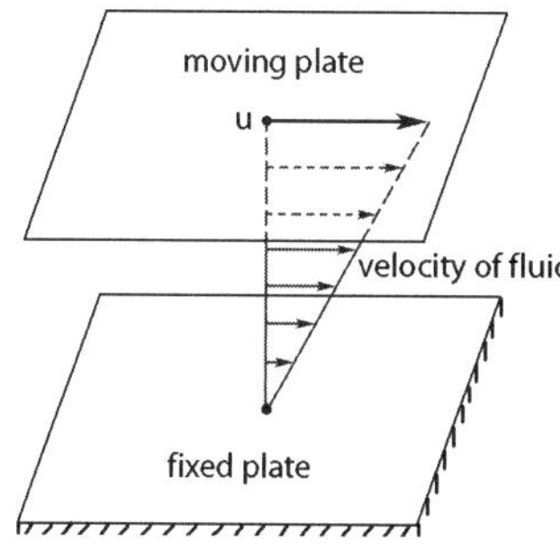

Chapter 3
Vector and Tensor Analysis

Mechanics is the paradise of the mathematical science, because in it we come to the fruit of mathematics.
L. da Vinci, *Notebook*, [68, p. 613].

Nothing puzzles me more than time and space; and yet nothing troubles me less, as I never think about them.
Charles Lamb, in a letter to Thomas Manning, January 2, 1806, in: R. Kaplan, *Science Says*, Freeman, New York, 2001, p. 77.

3.1 Scalars, Vectors, and Tensors

A tensor is something with subscripts.
M. Gunzburger, 1978, in response to a student's question: *What is a tensor?*

We are interested in domains in $\mathbb{R}^2$ and $\mathbb{R}^3$. To avoid useless repetition we are restricting our definitions to $\mathbb{R}^3$. In the case of $\mathbb{R}^2$ simply substitute "2" for "3" in the summations and definitions where appropriate. A collection of 3^n real numbers is an n-tensor. The first three examples are most interesting and are called scalars ($n = 0$ subscripts), vectors ($n = 1$ subscript), and tensors ($n = 2$ subscripts):

$$\begin{aligned}\text{0-tensor} &= \text{a real number} = \text{scalar},\\ \text{1-tensor} &= \text{3 real numbers} = \text{vector } \vec{v},\\ \text{2-tensor} &= 3 \times 3 \text{ array} = \text{tensor } \mathbf{T}.\end{aligned}$$

The classic examples are

$$\begin{aligned}\text{temperature, pressure} &= \text{scalar},\\ \text{force, velocity} &= \text{vector},\\ \text{deformation, stress} &= \text{tensor}.\end{aligned}$$

With the usual abuse of notation, we shall call a 2-tensor, a tensor.

The product, sum, transpose, trace, determinant, etc., of a (2-) tensor are defined exactly as for a 3×3 matrix. The vector dot product and tensor contraction are defined by

$$u \cdot v := \quad u_1 v_1 + u_2 v_2 + u_3 v_3 (= u_i v_i),$$
$$\mathbf{T} : \mathbf{S} := \sum_{i=1}^{3} \sum_{j=1}^{3} \mathbf{T}_{ij} \mathbf{S}_{ij} (= \mathbf{T}_{ij} \mathbf{S}_{ij}).$$

These definitions, and many calculations in vector and tensor analysis, can be compressed and simplified by the *summation convention* that

- indices repeated in a multiplicative term are summed from 1 to 3, and
- indices occurring after a comma denote differentiation with respect to that variable (e.g., $u_{,j} := \frac{\partial u}{\partial x_j}$).

With the summation convention,[16] the above dot products and contractions become

$$u \cdot v = u_i v_i,$$
$$\mathbf{T} : \mathbf{S} = \mathbf{T}_{ij} \mathbf{S}_{ij},$$

and,

$$u_i u_i = |u|^2, \; u_{,j} = \frac{\partial u}{\partial x_j} \text{ and } u_{,jj} = \sum_{j=1}^{3} \frac{\partial^2 u}{\partial x_j^2}.$$

With the summation convention, the definition of the tensor-vector product is consistent with matrix-vector multiplication

$$\mathbf{T} \cdot u := \mathbf{T}_{ij} u_j \text{ and } u \cdot \mathbf{T} := u_i \mathbf{T}_{ij}.$$

The typographic notation used to distinguish between scalars, vectors, and tensors is highly variable. After this point, vectors and scalars will be denoted by ordinary letters (no bold and no over arrows) and tensors by bold, capital letters. Every combination of notation is seen and no notation takes the place of understanding what the variables in an equation mean.

Let δ_{ij} denote the Kronecker δ:

$$\delta_{ij} = \begin{cases} 1 \text{ if } i = j, \\ 0 \text{ if } i \neq j. \end{cases}$$

The *identity tensor* **I** is defined as

$$\mathbf{I}_{ij} = \delta_{ij} = \begin{pmatrix} 1 & 0 & 0 \\ 0 & 1 & 0 \\ 0 & 0 & 1 \end{pmatrix}.$$

[16]This summation convention is very handy and makes calculations much simpler. For example, one of the exercises in the next section is to prove the vector identity $v \cdot \nabla v \cdot v = \frac{1}{2} \nabla \cdot (v|v|^2) - \frac{1}{2}|v|^2 \nabla \cdot v$. This is a very easy calculation using index notation and the summation convention but a much harder one using an index-free notation for differential operators.

Keeping the summation convention in mind, if u is a velocity field, $u : \Omega \subset R^3 \to R^3$, then the gradient tensor, ∇u, is the 3×3 array (i.e., 2-tensor) with ij entry

$$(\nabla u)_{ij} = \frac{\partial u_j}{\partial x_i} = u_{j,i}.$$

Remark 2. *Many find the definition of the tensor $(\nabla u)_{ij} = u_{j,i}$ visually jarring because of the subscript j, i on the right-hand side and the opposite ij on the left-hand side. However, this definition is necessary for vector-tensor products to be associative. For example, it is easy to check that $(u \cdot \nabla)v = u \cdot (\nabla v)$ precisely because of this definition of the gradient tensor and it fails if the subscript is i, j.*

The *deformation* and *spin tensors* [4], [87] associated with u are the symmetric and skew symmetric parts of the gradient (tensor) of u:

$$\mathbf{D}(u)_{ij} := \frac{1}{2}(\nabla u + \nabla u^t) = \frac{1}{2}(u_{j,i} + u_{i,j}),$$
$$\Omega(u)_{ij} := \frac{1}{2}(\nabla u - \nabla u^t) = \frac{1}{2}(u_{j,i} - u_{i,j}).$$

There is another good notation for the deformation and spin tensor that seems to be catching on in computational fluid dynamics. Since these are, respectively, the symmetric and skew-symmetric parts of the gradient tensor, they are increasingly written as[17]

$$\mathbf{D}(u) = \nabla^s u \text{ and } \Omega(u) = \nabla^{ss} u.$$

It is compact (although not insightful at all) to define the cross product of two vectors in terms using the permutation symbol $e(i, j, k)$. The permutation symbol is defined by

$$e(1, 2, 3) = e(2, 3, 1) = e(3, 1, 2) = 1,$$
$$e(2, 1, 3) = e(1, 3, 2) = e(3, 2, 1) = -1,$$
$$\text{all other } e(i, j, k) = 0.$$

Of course, no one computes cross products this way; everyone computes them using the formal expansion of the determinant formula that we have all seen in multivariable calculus.

Definition 13 (the cross product). *Let b, c be two vectors in $\mathbb{R}^3$. Then the ith component of the vector $b \times c$ is $(b \times c)_i := e(i, j, k) b_j c_k$.*

3.2 Vector and Tensor Calculus

The mathematical facts worthy of being studied are those which, by their analogy with other facts, are capable of leading us to the knowledge of a physical law. They reveal the kinship between other facts, long know, but wrongly believed to be strangers to one another.

H. Poincaré, in N. Rose, *Mathematical Maxims and Minims*, Rome Press, Raleigh, NC, 1988.

[17]These are good notations because they are almost self-explanatory. We will use both in this book at various places.

For scalar functions in $1-d$ there is only one type of derivative: $\frac{d}{dx}$. For vectors and tensors, there are several that are important: div, grad, and curl.

Definition 14 (div, grad, and curl). *If $v : \Omega \to \mathbb{R}^3$, then* div $v = \nabla \cdot v$ *is the scalar defined by*

$$\nabla \cdot v = \frac{\partial v_1}{\partial x_i} + \frac{\partial v_2}{\partial x_2} + \frac{\partial v_3}{\partial x_3} = v_{i,i}.$$

If $f : \Omega \to R$, then $\mathrm{grad}(f) = \nabla f$ *is the vector whose* ith *component is*

$$(\nabla f)_i = \frac{\partial f}{\partial x_i}.$$

If $v : \Omega \to \mathbb{R}^3$ is a vector and $e(\cdot,\cdot,\cdot)$ is the permutation symbol, then curl $v = \nabla \times v$ *is the vector defined*[18] *whose* ith *component is*

$$(\nabla \times v)_i = e(i, j, k) v_{k,j}\,.$$

All three differential operators are important in fluid mechanics. The divergence of a vector field measures the local flux per unit volume. The curl measures the vector field's local circulation per unit area.

Definition 15 (vorticity). *Let u denote the velocity of a fluid motion. Then*

$$\omega = \nabla \times u$$

is the fluids vorticity.

Example 5 (vorticity in shear flows). The vorticity represents the turning force the flow exerts. Thus it is connected with the swirly-ness of the flow but it is not exactly this. For example, if a fluid is enclosed between two glass plates one unit length apart and the top plate is moved with uniform velocity one, a shear flow results;[19] see Figure 3.1.

It is traditional in studying this flow to let the x axis point along in the flow direction, the y axis point up, and the z axis point in the third direction (in which not much is happening). It can be shown that a linear velocity interpolating between the bottom plates zero velocity and the top plates unit velocity is a solution of the Navier–Stokes equations governing the shear flow problem. This gives the velocity

$$u(x, y, z) = (u_1, u_2, u_3)(x, y, z) = (y, 0, 0).$$

[18]This is the mathematically correct definition but, of course, everyone actually calculates the curl using the memory-aid formula

$$\nabla \times v = \det \begin{pmatrix} \widehat{i} & \widehat{j} & \widehat{k} \\ \frac{\partial}{\partial x_1} & \frac{\partial}{\partial x_2} & \frac{\partial}{\partial x_3} \\ v_1 & v_2 & v_3 \end{pmatrix}.$$

[19]If the plates are not infinite in one horizontal direction, say, the x direction, but finite and if periodic boundary conditions are imposed in that direction, the flow problem becomes very close to flow between rotating cylinders. Flow between rotating cylinders is one of the classic problems in experimental fluid dynamics and is described in Chapter 5.

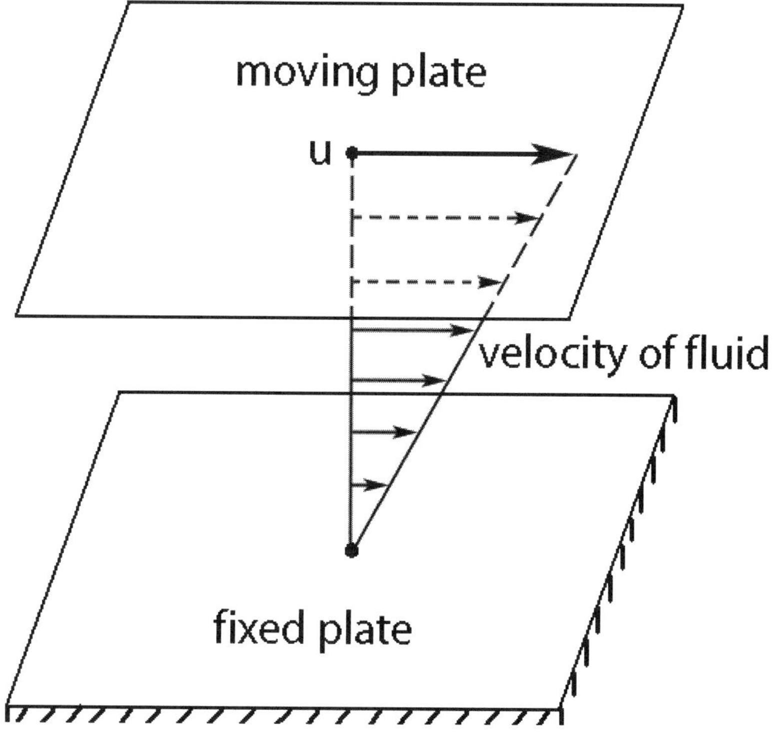

Figure 3.1. *Geometry of a simple shear flow.*

Imagine placing a little (infinitesimal) paddle wheel in the flow so it can spin in the $x - y$ plane.[20] The flow just above the wheel is faster than that just below. Thus, although the flow is not swirly, the wheel will spin and, consequently, $\nabla \times u \neq 0$![21] Interestingly, this linear shear flow velocity is not always the observed flow in experiments. This flow becomes turbulent if a measure of the turning force the fluid exerts on itself becomes large enough. One heuristic description of why this happens is that when the fluid's vorticity becomes large enough, it turns the flow itself and the flow then becomes swirly. ∎

Proposition 4. *Let u denote the velocity field of a fluid and let Ω be the flow domain on which the velocity is defined. Assume $u \in C^2(\Omega)$. Then, for any $x \in \Omega$ and any displacement vector h,*

$$u(x + h, t) = u(x, t) + (\nabla u) \cdot h + O(|h|^2)$$

and

$$u(x + h, t) = u(x, t) + \mathbf{D} \cdot h + \frac{1}{2}\omega \times h + O(|h|^2). \tag{3.2.1}$$

[20]If v denotes the fluid velocity, then $\nabla \times v$ has the interpretation as a local turning force. Specifically, let this paddle wheel spin about an axis denoted by the unit vector $\widehat{n}$ (so it spins in the plane perpendicular to $\widehat{n}$). If $(\nabla \times v) \cdot \widehat{n} = 0$, then the wheel will stay at rest. If $(\nabla \times v) \cdot \widehat{n} > 0$, then the wheel will spin counterclockwise, and if $(\nabla \times v) \cdot \widehat{n} < 0$, then the wheel will spin clockwise.

[21]You should not believe this just because it is written down. Calculate $\nabla \times u$ and check that it is nonzero!

Proof. This is an exercise in Taylor series! □

This expansion (3.2.1) shows that as we move from one point (x, t) to a nearby point, the fluid velocity is altered by a rigid body rotation $(h \times \omega/2)$ and a deformation $(\mathbf{D} \cdot h)$. This explains the interpretation

D = local velocity deformation and

ω = **local rate of rotation/turning force**.

The three big theorems of vector analysis (Green's, Stokes's, and the divergence theorem) are all important in mathematical fluid mechanics. We state the divergence theorem here to emphasize its centrality.

Theorem 10 (the divergence theorem). *Let Ω be a domain in $\mathbb{R}^3$. Suppose $\partial\Omega$ is Lipschitz continuous. Then, for any function $f \in C^1(\Omega)$*

$$\int_\Omega \frac{\partial f}{\partial x_i} dx = \int_{\partial\Omega} f n_i d\sigma, \ i = 1, 2, \text{ and } 3,$$

where $\hat{n} = (n_1, n_2, n_3)$ is the outward unit normal to $\partial\Omega$.

Proof. *See, for example, Aris* [4] *or Schey* [85]. □

In particular, the divergence theorem implies that for any vector function $v : \Omega \to \mathbb{R}^3$

$$\int_\Omega \nabla \cdot v dx = \int_{\partial\Omega} v \cdot \hat{n} d\sigma.$$

Remark 3. *So far everything presented for $\Omega \subset \mathbb{R}^3$ holds for two-dimensional domains $\Omega \subset \mathbb{R}^2$ with vector functions $v : \Omega \to \mathbb{R}^2$ and 2-tensors (being 2×2 arrays in $\mathbb{R}^2$).*

Definition 16 (potential, incompressible, and irrotational flows). *Let v denote the velocity field of a fluid motion. The flow field is* potential *if*

$$v = \nabla\phi$$

for some scalar function ϕ, irrotational *or curl-free if*

$$\omega = \nabla \times v = 0$$

and incompressible *or div-free if*

$$\nabla \cdot v = 0.$$

It is known that if Ω is simply connected (e.g., has no holes in two dimensions), then v is irrotational if and only if it is potential. In fact, a lot more is known about these concepts.

Definition 17 (connectedness). *A bounded open domain Ω is* path-connected *(also called 0-connected) if each pair of points in Ω can be joined by a continuous curve inside Ω.*

Ω *is* simply connected *(also called* 1*-connected) if it is path connected and each closed loop inside* Ω *can be shrunk continuously to a point staying inside* Ω*.*

Ω *is* 2-connected *if it is* 1*-connected and each sphere inside* Ω *can be shrunk continuously to a point staying inside* Ω*.*

Theorem 11 (scalar and vector potentials). *Let* $\Omega \subset \mathbb{R}^d$ *be a bounded open set and let* $v : \Omega \to \mathbb{R}^d$ *be a smooth function.*

(i) *If* Ω *is simply connected and* $\nabla \times v = 0$*, then* $v = \nabla\phi$ *for some scalar function* ϕ*.*

(ii) *If* Ω *is* 2*-connected and* $\nabla \cdot v = 0$*, then there is a vector function* ψ *with* $v = \nabla \times \psi$*.*

The Helmholtz decomposition theorem and its generalization, known as the Hodge theorem, gives a global codification of the above picture. It, and its generalization to function spaces, is fundamental to analysis of vector functions.

Definition 18. *The space of div-free functions is denoted*

$$H := \{u \in C^1(\Omega) : \nabla \cdot u = 0 \text{ in } \Omega \text{ and } u \cdot \hat{n} = 0 \text{ on } \partial\Omega\}$$

and the space of potential functions is denoted

$$K := \{v \in C^1(\Omega) : v = \nabla\phi \text{ for some scalar function } \phi\}.$$

Theorem 12 (Helmholtz decomposition theorem). *Suppose* $\Omega \subset \mathbb{R}^3$ *is a* 2*-connected domain with smooth enough boundary. Any* $C^1(\Omega)$ *vector function* $w : \Omega \to \mathbb{R}^3$ *can be uniquely written as*

$$w = u + v, \quad u \in H, \quad v \in K.$$

Furthermore, for any $u \in H,\ v \in K$

$$\int_\Omega u \cdot v \, dx = 0.$$

3.3 Conservation Laws

Nothing is permanent; all is flux.
Heraclitus of Ephesus (535–475 BC), in *Cratylus* by Plato.

The divergence theorem gives a powerful tool for developing equations describing physical systems in which "something" is in flux in such a way that it is conserved. Define the key variables:

$$\begin{aligned} \phi(x,t) &:= \text{local density of conserved quantity,} \\ q(x,t) &:= \text{local flux (vector) of conserved quantity,} \\ f(x,t) &:= \text{local sources or sinks per unit volume.} \end{aligned}$$

Let B_ϵ be a ball about x of radius ϵ. Conservation implies

$$\begin{aligned} &\text{Rate of change of net amount in } B_\epsilon \\ &\qquad = \text{net flux through } \partial B_\epsilon + \text{sources/sinks} * \text{volume of } B_\epsilon, \end{aligned}$$

or, in equations (since $\hat{n}$ is the outward normal to B_ϵ),

$$\frac{d}{dt}\int_{B_\epsilon}\phi(x,t)dx = -\int_{\partial B_\epsilon} q\cdot\hat{n}d\sigma + \int_{B_\epsilon} f(x,t)dx.$$

Applying the divergence theorem, dividing by $vol(B_\epsilon)$, and shrinking B_ϵ to a point gives

$$\frac{\partial\phi}{\partial t} + \nabla\cdot q = f(x,t), \tag{3.3.1}$$

This must be supplemented with an equation relating to the density ϕ to the flux q:

$$q = Q(\phi). \tag{3.3.2}$$

This should properly be considered to be an experimental law whose exact form depends on the nature of the "something" which is being conserved. Combining (3.3.1) and (3.3.2) gives the equation

$$\frac{\partial\phi}{\partial t} + \nabla\cdot Q(\phi) = f(x,t).$$

Example 6 (conservation of mass in a fluid). A fluid convects its own particles. Letting $\rho(x,t)$ be the fluid density and $u(x,t)$ be the fluid velocity, the mass flux is simply $q = \rho u$. Thus, the fluid density satisfies

$$\frac{\partial}{\partial t}\rho + \nabla\cdot(\rho u) = 0.$$

If the fluid is homogeneous[22] and *incompressible*, $\rho(x,t) \equiv \rho_0$, and this reduces to

$$\nabla\cdot u = 0.$$

Thus, constant density imposes a constraint on the velocity field of a fluid.

This is the first equation of mathematical fluid dynamics. It is called the *continuity equation* because it arises from shrinking a volume to a point—a process requiring continuity of all variables. ∎

Example 7 (conservation of momentum). Since a liquid convects itself, it also convects its own momentum. Thus, with a small adaptation, the conservation law framework also gives the equations describing conservation of momentum in a fluid. Let ρ be the fluid density (assumed constant here) and $u(x,t)$ be the fluid velocity; the momentum vector is simply $m(x,t) = \rho u$. The general form of conservation laws tells us that if the resulting equation is to be dimensionally correct (in the simplest sense of the number of subscripts agreeing in each term), the momentum flux must be a tensor. Also, it is clear that sources of local momentum must be included in the right-hand side. We thus have the variables

$$\begin{aligned} m(x,t) &= \text{momentum density (vector)} = \rho u,\\ \mathbf{Q} &= \text{momentum flux (tensor)} = um = \rho u_i u_j,\\ S &= \text{momentum sources} = \text{force (vector)}. \end{aligned}$$

[22]Homogeneous ≈ Composed of one fluid. It is possible to have (incompressible) mixtures of two incompressible fluids whose density varies due to the mixing proportions varying as the fluids flow.

Conservation of momentum gives

$$\frac{\partial}{\partial t}(\rho u) + \nabla \cdot \mathbf{Q} = S \text{ or}$$
$$\frac{\partial}{\partial t}(\rho u) + \nabla \cdot (\rho u u) = S.$$

We compute, using $\nabla \cdot u = 0$,

$$\nabla \cdot (uu) = (u_i u_j)_{,i} = u_{i,i} u_j + u_i u_{j,i} = 0 + u \cdot \nabla u.$$

Thus, for an incompressible fluid, the equation for conservation of momentum becomes

$$\frac{\partial}{\partial t} u + u \cdot \nabla u = \frac{1}{\rho_0} S.$$

The momentum sources in a volume are caused by two effects: body forces, such as gravity and Coriolis forces in a rotating fluid, and the internal drag exerted on a volume containing fluid by the fluid adjacent to it. Let $f(x,t)$ collect all the body forces. Developing a model of the internal forces in a fluid is nontrivial and will be done in detail in Chapter 5. For now, it suffices to say that these are accounted for by a tensor Π called the stress tensor,and the final result is

$$\frac{\partial}{\partial t} u + u \cdot \nabla u = \frac{1}{\rho_0}(\nabla \cdot \Pi + \vec{f}),$$

which is called the momentum equation. ∎

Example 8 (transport of a passive scalar). A "passive scalar" includes pollutants, dust particles, or even heat and is transported by fluid with (assumed known) velocity field $u(x,t)$. In this case,

$$\phi(x,t) := \text{ local density of passive scalar}$$

$$\begin{aligned} q(x,t) &:= \text{ flux of passive scalar} \\ &= \text{ molecular diffusion in fluid and convection by flow.} \end{aligned}$$

Molecular diffusion obeys the analogue of Fourier's law of heat conduction, known as *Fick's law of diffusion*:

$$\begin{aligned} \text{flux due to molecular diffusion } &\propto -\nabla \phi \text{ and} \\ \text{flux due to convection } &= u\phi. \end{aligned}$$

Calling μ the proportionality constant gives

$$q = -\mu \nabla \phi + u\phi.$$

This gives the conservation law known as the convection-diffusion equation:

$$\frac{\partial \phi}{\partial t} + \nabla \cdot (-\mu \nabla \phi + u\phi) = f.$$

If diffusion is negligible, we get

$$\frac{\partial \phi}{\partial t} + \nabla \cdot (u\phi) = f,$$

while if convection is negligible, we get

$$\frac{\partial \phi}{\partial t} - \nabla \cdot (\mu \nabla \phi) = f. \quad \blacksquare$$

Example 9 (conservation of energy). If $\phi(x, t)$ is the heat density, then

$$\begin{aligned} \phi(x, y) &= \rho c_p T(x, t), \\ T(x, t) &= \text{ temperature}, \\ \rho &= \text{density of material}, \\ c_p &= \text{ specific heat} \\ r &= \text{ external heat sources}. \end{aligned}$$

The heat flux q contains two contributions. The first is given by Fourier's law of heat conduction as $-k\nabla T$ for the conduction of heat. The second is convection of heat whose contribution to the flux is written $u\, T$. Adding these two effects gives

$$q = -k\nabla T + u\phi.$$

The total heat source also has two contributions. The first is simply the external heat source r. This can be due to radiation, for example. The second corresponds to heat caused by the work done by the fluid (in essence) getting out of its own way. Handwaving a bit, this is written as $\Pi : (\nabla u)$, where Π is the stress tensor, introduced above and studied in detail in Chapter 5. This yields

$$\frac{\partial}{\partial t}(\rho c_p T) + \nabla \cdot (-k\nabla T + u\rho c_p T) = r + \Pi : (\nabla u).$$

The second contribution to the heat source, $\Pi : (\nabla u)$, is negligible in some important applications.[23] $\blacksquare$

Example 10 (saturated underground flow). Consider a region underground which is saturated with a fluid which is flowing through its pores. To develop the conservation law describing this flow, introduce the variables:

$$\begin{aligned} \phi &:= \text{porosity or fraction of volume available for the fluid}, \\ \rho &:= \text{ fluid density}, \\ f &:= \text{ mass source or sink (e.g., a well)}, \\ q &:= \text{ flux = flow rate through a surface per unit area per unit time}. \end{aligned}$$

[23] Since the term $\nabla \cdot (\Pi u)$ introduces much greater mathematical difficulties, there are many other applications in which it is optimistically presumed to be negligible.

Given a volume V we can calculate

$$\text{total mass in } V = \int_V \rho\phi \, dx,$$
$$\text{mass flux through } \partial V = -\int_{\partial V} \rho q \cdot \hat{n} \, ds,$$

where $\hat{n}$ is the outward unit normal to ∂V. Conservation of mass implies, as before, for any V,

$$\frac{d}{dt}\int_V \rho\phi \, dv = -\int_{\partial V} \rho q \cdot \hat{n} \, ds + \int_V f \, dx.$$

Shrinking V to a point gives the conservation law

$$(\rho\phi)_t + \nabla \cdot (\rho q) = f.$$

To close the system we need a relation for $\vec{q}$. In porous media, this relation is known as Darcy's law. The French engineer H. P. G. Darcy[24] discovered that by doing simple "bucket tests" of water under various pressures percolating through sand, the flow rate $\vec{q}$ was proportional to the pressure difference driving the flow.

Darcy's law states that the flux is proportional to the pressure gradient, or

$$q = -\left(\frac{k}{\mu}\right)(\nabla p + \rho g),$$

where g is the gravitational force vector, μ is the fluids viscosity, and the proportionality constant k is called

$$k := \text{material permeability}.$$

Darcy's law gives the systems

$$(\rho\phi)_t = -\nabla \cdot \left[\rho\left(\frac{-k}{\mu}\right)[\nabla p + \rho g]\right] + f.$$

If the fluid is a gas, then an equation of state relates ρ to p and closes the system.

If the fluid is an incompressible liquid, then

$$\rho \doteq \rho_0 = \text{constant}.$$

If additionally $\phi = \phi(x)$ so the problem has reached steady state, we then have

$$-\nabla \cdot \left(\left(\frac{\rho k(x)}{\mu}\right)\nabla p\right) = \widetilde{f}, \text{ where } \widetilde{f} = \nabla \cdot \left(\frac{\rho^2 k(x,y) g}{\mu}\right) + f,$$

which is a Poisson problem for the pressure $p(x)$.

The most natural boundary conditions are either a known pressure or flux:

$$p = p_0 \text{ on part of the boundary,}$$
$$\frac{k}{\mu}\nabla p \cdot \hat{n} + \rho_0 g \cdot \hat{n} = r(x), \text{ on the rest of the boundary.} \quad \blacksquare$$

[24]H. P. G. Darcy (1803–1858) was a French engineer who made significant contributions to pipe flow (the Darcy–Weisbach equation), guessed the existence of boundary layers, and conducted bucket tests that discovered the remarkably simple Darcy law for flow in a porous medium.

Example 11 (conservation of energy in more detail). The above development of the energy equation for conservation of energy involved some handwaving at a critical step. This is because the thermal energy in a flow is not, in itself, conserved. The total energy, made up of kinetic energy plus thermal energy, is the appropriate conserved quantity. Next we rederive the energy equation from conservation of total energy.

For an incompressible fluid ρ is constant so conservation of mass reads

$$\nabla \cdot u = 0,$$

while conservation of momentum is given by

$$\rho\,(u_t + u \cdot \nabla u) - \nabla \cdot \Pi = f.$$

Let

$$\begin{aligned} \phi &= \rho(c_p T + |u|^2/2), \\ q &= k\nabla T + \Pi u, \\ s &= f \cdot u + r. \end{aligned}$$

Here ϕ is the energy per unit volume of a fluid element, q is the flux through the control volume boundary, and s collects the sources inside the control volume. Πu represents the work done by the traction, $(\Pi u) \cdot \widehat{n} = (\Pi^{tr}\widehat{n}) \cdot u$, and $f \cdot u$ is the work done by the interaction of the flow with body forces. r represents the heat sources due to radiation and other, external sources. Inserting this into the conservation law

$$\phi_t + \nabla \cdot (\phi u - q) = s$$

gives

$$\left(\rho(c_p T + |u|^2/2)\right)_t + \nabla \cdot \left(\rho(c_p T + |u|^2/2)u - K\nabla T - \Pi u\right) = f \cdot v + r.$$

When ρ is constant and $\mathrm{div}(u) = 0$, it is elementary to show that the kinetic energy itself satisfies

$$\begin{aligned} (\rho|u|^2/2)_t + \nabla \cdot \left((\rho|u|^2/2)u - \Pi u\right) &= (\rho(u_t + u \cdot \nabla u) - \nabla \cdot (\Pi)) \cdot u - \Pi : (\nabla u) \\ &= f \cdot u - \Pi : (\nabla u), \end{aligned}$$

where the second inequality follows from the momentum equation. Subtracting the kinetic energy equation from the total energy equation, it follows that

$$(\rho c_p T)_t + \nabla \cdot \left(\rho c_p T u - k\nabla T\right) = r + \Pi : (\nabla u). \quad \blacksquare$$

3.4 Remarks on Chapter 3

For more details on vectors and tensors, see Schey [85] and Aris [4]. For more details on vorticity in fluid motion, see [75] (a historically interesting book) and the first section of Serrin [87]. Serrin's article is famous in mathematical fluid dynamics. It is both mathematically beautiful and physically lucid. The mathematics of conservation laws is developed in the book by Whitham [102]. The conservation law idea is central to the development of the NSE and, if possible, yet more important in compressible flows. Although we live in three dimensions, two-dimensional flows are still important. A two-dimensional vector field $(v_1(x, y), v_2(x, y))$ is properly interpreted as a three-dimensional field that is uniform in the third dimension, $(v_1(x, y), v_2(x, y), 0)$. Vector fields in two dimensions give accurate descriptions of flows over long, thin bodies (such as airfoils), thin films, and many atmospheric flows (which are largely two-dimensional flows punctuated by localized and sometimes intense three-dimensional effects).

3.5 Exercises

Exercise 24. *Find a good book on vector analysis, for example, Aris* [4] *or Schey* [85], *and write a short synopsis of the geometric interpretation of the dot and cross product.*

Exercise 25. *Show that $D : \Omega = 0$ because $A : B = 0$ whenever A is symmetric and B is skew-symmetric.*

Exercise 26 (two vector identities). *Using subscript notation, show that for any smooth vector function*

$$v \cdot \nabla v \cdot v = \frac{1}{2}\nabla \cdot (v|v|^2) - \frac{1}{2}|v|^2\nabla \cdot v$$

and that if $\nabla \cdot u = 0$, then

$$\nabla \cdot (u \cdot \nabla u) = \nabla u : \nabla u^t.$$

Exercise 27. *Prove that $\Omega \cdot h = \omega \times h/2$ and use this and Taylor's theorem to prove Proposition* 4.

Exercise 28. *Suppose $u : \Omega \subset \mathbb{R}^3 \to \mathbb{R}^3$ is a smooth function on a polygonal domain Ω and satisfies*

$$\nabla \cdot u = 0 \;\; and \;\; u = 0 \;\; on \;\; \partial\Omega.$$

Show, using the divergence theorem, that

$$\begin{aligned} -\int_\Omega (\Delta u) \cdot u \, dx &= \int_\Omega \nabla u : \nabla u \, dx \\ &= 2\int_\Omega \mathbf{D}(u) : \mathbf{D}(u) \, dx = \int_\Omega |\nabla \times u|^2 dx. \end{aligned}$$

The quantity $\int_\Omega |\nabla \times u|^2 dx$ is called the enstrophy.

Exercise 29 (two-dimensional vector analysis). *Vector functions $u : \Omega \subset R^2 \to R^2$ can be extended to vector functions on the cylinder*

$$u(x, y, z) : \Omega \times (-\infty, \infty) \subset \mathbb{R}^3 \to \mathbb{R}^3$$

by $u = (u_1, u_2, 0)$. Give a definition of div *and* curl *for such functions, consistent with the three-dimensional definition. Check yours against Aris* [4] *or Schey* [85].

Exercise 30. *Show that* $\nabla \cdot \nabla v = \nabla\nabla \cdot v - \nabla \times \nabla \times v$ *(i.e.,* div grad v = grad div v − curl curl v*).*

Exercise 31. *Show that all potential flows are irrotational. Next, either prove or find a proof that all irrotational flows in a simply connected domain are potential.*

Exercise 32 (a curl-free field that is not potential). *Let*

$$D := \{(x, y) : x^2 + y^2 > 1\}$$

$$\text{and } v = (-y/(x^2 + y^2), x/(x^2 + y^2)).$$

Graph a few velocity vectors to check that the velocity rotates about the origin. Show $\nabla \times v = 0$ *but* v *is not potential. This is an example that shows that connectedness cannot be dropped. In the punctured domain, this* v *is irrotational but not potential!*

Exercise 33. *Show that* $||\nabla \cdot u|| \leq C||\nabla u||$ *and find the constant* C *in terms of the dimension* $d = 2$ *or* 3.

Exercise 34 (a difference approximation to $\frac{\partial}{\partial t}\phi + u \cdot \nabla\phi$). *Given time levels* t_{old}, $t_{new} = t_{old} + \Delta t$, *and a point* P *in the flow domain, let* Q *denote the point* $P - u(t_{old}, P)\Delta t$. *Give an interpretation of the point* Q *relative to* P. *Show by a Taylor series expansion that the following is a first order approximation to the material derivative:*

$$\left(\frac{\partial}{\partial t}\phi + u \cdot \nabla\phi\right)(t_{new}, P) = \frac{\phi(t_{new}, P) - \phi(t_{old}, Q)}{\Delta t} + O(\Delta t).$$

Give a graphical interpretation of the difference quotient.

Exercise 35 (the kinetic energy equation). *Begin with the equation of conservation of momentum:*

$$\rho\,(u_t + (u \cdot \nabla)u) - \nabla \cdot \Pi = f.$$

For constant ρ *and* $\nabla \cdot u = 0$, *take the dot product of the momentum equation and rearrange to show that the kinetic energy,* $\rho|u|^2/2$, *satisfies*

$$\begin{aligned}(\rho|u|^2/2)_t + \nabla \cdot \left((\rho|u|^2/2)u - \Pi u\right) &= (\rho(u_t + (u \cdot \nabla)u) - \nabla \cdot (\Pi)) \cdot u - \Pi : (\nabla u)\\ &= f \cdot u - \Pi : (\nabla u).\end{aligned}$$

Part II

Steady Fluid Flow Phenomena

Chapter 4

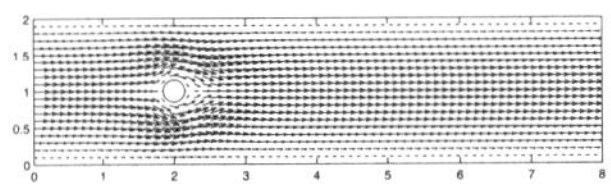

Approximating Vector Functions

The finite element method has been an astonishing success. It was created to solve the complicated equations of elasticity and structural mechanics and, for these problems, it has essentially superseded the method of finite differences. Now, other applications are developing.

G. Strang and G. Fix, preface to [92], 1973.

4.1 Introduction to Mixed Methods for Creeping Flow

Many times the problems we wish to solve are inherently vector valued since they deal in forces, velocities, and flux. As one example, reconsider the steady porous media flow problem of finding the pressure p driving the flow in the media. The pressure satisfies

$$-\nabla \cdot (k(x, y)\nabla p) = f \text{ in } \Omega, \quad p = 0 \text{ on } \partial\Omega.$$

The physically interesting variable is not the velocity potential p but rather the flux or Darcy velocity $u = -k\nabla p$. This leads to the problem of finding (u, p) satisfying

$$\begin{aligned} \nabla \cdot u &= f \text{ and } k^{-1}u + \nabla p = 0 \text{ in } \Omega \\ p &= 0 \text{ on } \partial\Omega. \end{aligned}$$

The Stokes problem of fluid mechanics, which will follow as a special case of the general equations of fluid motion in a later chapter, describes creeping flow. It has a similar mathematical structure to the porous media equation. The Stokes problem consists of finding the fluid velocity $u : \Omega \to \mathbb{R}^d$ and pressure $p : \Omega \to \mathbb{R}$ defined in the flow domain Ω[25] satisfying

$$\begin{aligned} &-\Delta u + \nabla p = f \text{ in } \Omega \subset \mathbb{R}^d \ (d = 2, \text{ or } 3), \\ &\nabla \cdot u = 0 \ \text{ in } \Omega \ \text{ and } u = 0 \text{ on } \partial\Omega. \end{aligned} \tag{4.1.1}$$

[25] We shall always assume that the flow domain is a bounded domain with, at least, a Lipschitz continuous boundary.

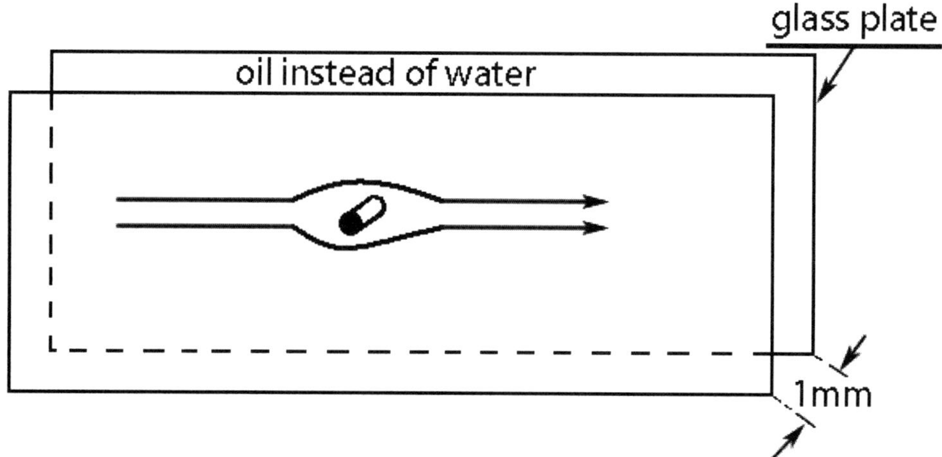

Figure 4.1. *Typical experimental realization of creeping flow.*

Note that ∇p occurs in the Stokes problem rather than p and that there is (under the above most common boundary conditions at least) no pressure boundary condition. Thus, the pressure can, at best, be determined only up to an additive constant ($\nabla(p + C) = \nabla p$). For this reason the pressure is normalized in some way to determine the arbitrary additive constant; the most mathematically convenient way is by $\int_\Omega p dx = 0$. In this chapter we focus on FEMs for the Stokes problem (4.1.1). There is also a general theory for mixed methods and a parallel theory for the porous media problem.

The primitive or physical variables for the porous media problem and the Stokes problem are u and p, velocity and pressure. There are an infinite number of changes of variables possible so there are vary many other possible combinations of variables (and which can be tested for their numerical solution). One feature of mixed methods is that they solve the physical problem (porous media or Stokes) in the physical variables. If our physical understanding is correct and if our mathematical representation of this understanding is correct, then mixed methods must be a good computational approach. When mixed methods were first tested (and for some years thereafter), problems, now corrected, were found that were traced back to insufficient understanding. The mathematical theory developed, presented in this chapter, has proved itself to be essential for stable and accurate solution of these (and other) problems in the most important and most physical variables.

Example 12 (creeping flow). Creeping flow refers to a solution of the Stokes problem. Although it is not the generic flow around us, there are many important flows governed by the Stokes problem. Typically, creeping flows occur in three settings: small geometries such as in human capillaries and flows in small pores such as in filtration, small velocities (hence "creeping"), and large viscosities as in casting flows and flows of lubricants. Physical experiments of creeping flows are done by combining all three factors. For example, one way to visualize a Stokes flow is to enclose a very viscous fluid (such as an oil) between parallel glass plates with a very small gap, typically O(1 mm), as in Figure 4.1.

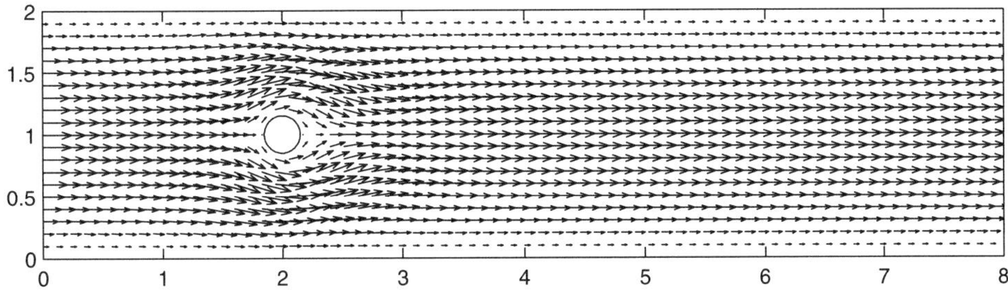

Figure 4.2. *Velocity vectors for Stokes flow.*

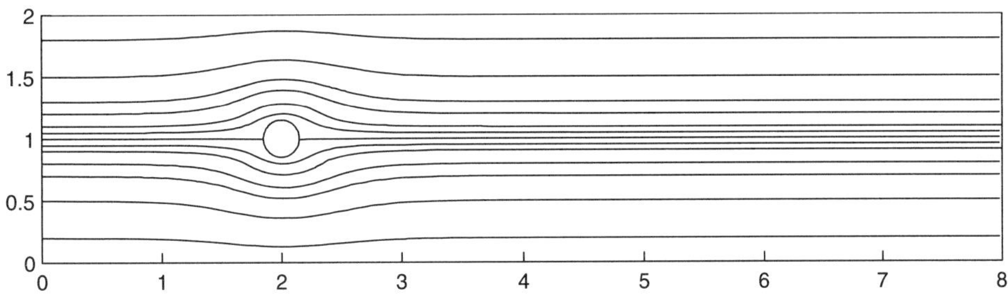

Figure 4.3. *Streamlines for Stokes flow.*

Solutions to the Stokes problem are typically quite smooth and the streamlines quite regular. Figures 4.2 and 4.3 are an example of the velocity field and streamlines[26] of such a Stokes flow. ■

First, we derive the variational formulation of the Stokes problem (4.1.1). Let (u, p) be a classical solution of the Stokes problem. Multiply (4.1.1) by (smooth enough) functions (v, q) and integrate:

$$\int_\Omega -\Delta u \cdot v + \nabla p \cdot v \, dx = \int_\Omega f \cdot v \, dx, \int_\Omega \nabla \cdot u \, q \, dx = 0.$$

Applying the divergence theorem, term by term, gives

$$\begin{aligned} -\int_\Omega \Delta u \cdot v \, dx &= -\int_\Omega u_{i,jj} v_i dx = -\int_\Omega (u_{i,j} v_i)_{,j} - u_{i,j} v_{i,j} dx \\ &= \int_\Omega \nabla u : \nabla v \, dx - \int_{\partial\Omega} \nabla u \cdot v \cdot \hat{n} ds. \end{aligned}$$

If v vanishes on $\partial\Omega$ we thus have

$$\int_\Omega -\Delta u \cdot v \, dx = \int_\Omega \nabla u : \nabla v \, dx.$$

[26]These two figures reveal another interesting effect. Plotting velocity vectors often loses the key solution behavior in a forest of little arrows. Flow visualization often involves a search for some scalar function of velocity (to be plotted instead) that can reveal the important flow behavior.

Similarly, if v vanishes on $\partial\Omega$,

$$\int_\Omega \nabla p \cdot v \, dx = \int_\Omega -p\nabla \cdot v \, dx.$$

Thus, for all v vanishing on $\partial\Omega$ and smooth enough and all q smooth enough, (u, p) satisfies

$$\int_\Omega \nabla u : \nabla v - p\nabla \cdot v \, dx = \int_\Omega f \cdot v \, dx, \quad \int_\Omega q\nabla \cdot u \, dx = 0. \tag{4.1.2}$$

The equation (4.1.2) points the way to the correct variational formulation of the Stokes problem (4.1.1). Define the norm

$$||v||_X := [||v||^2 + ||\nabla v||^2]^{1/2},$$

where (recall that v is a vector and ∇v a tensor), as usual,

$$||v||^2 = \sum_i ||v_i||^2, \; ||\nabla v||^2 = \sum_{i,j} \left\| \frac{\partial v_i}{\partial x_j} \right\|^2.$$

The velocity space X is defined (e.g., in Chapter 1) as

$$X = \text{ closure in } || \cdot ||_X \text{ of}$$
$$\left\{ v \in C^1(\Omega) : v_i \in L^2(\Omega), \; \frac{\partial v_j}{\partial x_i} \in L^2(\Omega) \; \forall \, i\,, j, \text{ and } v = 0 \text{ on } \partial\Omega \right\}.$$

The pressure space Q does not require any differentiability since no derivatives of p or q appear in (4.1.2). On the other hand, it must account for the fact that if (u, p) is a solution of (4.1.1), then so is $(u, p + c)$ for any constant c. Accordingly, to fix the value of the undetermined, additive constant it is usual to impose the condition of mean value zero and thus define

$$Q := \left\{ q \in L^2(\Omega) : \int_\Omega q \, dx = 0 \right\} \text{ (also denoted } L_0^2(\Omega)).$$

We thus come to the variational formulation of (1.1): find $u \in X$, $p \in Q$ satisfying

$$\begin{aligned} &\nu(\nabla u, \nabla v) - (p, \nabla \cdot v) = (f, v) \; \forall \, v \in X, \\ &(q, \nabla \cdot u) = 0 \; \forall \, q \in Q. \end{aligned} \tag{4.1.3}$$

4.2 Variational Formulation of the Stokes Problem

In reflecting on the principles according to which the motion of a fluid ought to be calculated when account is taken of the tangential force, and consequently. . . .

G. Stokes, in *On the Theories of the Internal Friction of Fluids in Motion*, 1845.

The fundamental variational formulation of the Stokes problem is (4.1.3) above:

$$\text{find } (u, p) \in (X, Q) = Y \text{ satisfying}$$
$$(\nabla u, \nabla v) - (p, \nabla \cdot v) = (f, v) \text{ and } (q, \nabla \cdot u) = 0,$$
$$\forall (v, q) \in (X, Q) = Y.$$

Lemma 5 (continuity). *$b(v, q) = \int_\Omega q\nabla \cdot v dx$ is continuous on $X \times Q$:*

$$|b(v, q)| \leq C||q||_Q||v||_X.$$

As a consequence, the divergence free subspace V *of X*

$$V = \{v \in X : (q, \nabla \cdot v) = 0 \ \ \forall q \in Q\}$$

is a closed subspace of X.

Proof. The inequality is obvious. Since $b(v, q)$ is bilinear it follows equally easily that V is a subspace. The key to showing V is closed is to use continuity. By continuity, let $v_n \in V$ be such that $v_n \to v$ in X and let $q \in Q$ be fixed; then $b(v_n, q) \to b(v, q)$. Since $b(v_n, q) \equiv 0$ it follows from continuity that $b(v, q) = 0$, i.e., $v \in V$. $\square$

Definition 19. *For $f \in L^2(\Omega)$, the H^{-1} norm and the V^* norm of f are*

$$||f||_{-1} := \sup_{v \in X} \frac{(f, v)}{||\nabla v||}, \ ||f||_* := \sup_{v \in V} \frac{(f, v)}{||\nabla v||}.$$

Definition 20. *The function spaces $H^{-1}(\Omega)$, and V^* are, respectively, the closures of $L^2(\Omega)$ in, respectively, $||\cdot||_{-1}$ and $||\cdot||_*$.*

Operationally, the *dual norm*[27] $=$ *norm on the dual of* $V = V^*$ norm of a function, $||f||_*$, is the smallest norm satisfying

$$(f, v) \leq ||f||_*||\nabla v||\forall v \in V,$$

and the H^{-1}norm = norm on the dual of $H_0^1(\Omega) = (H_0^1(\Omega))^*$ norm is the smallest satisfying this upper bound for the wider collection of functions of any $v \in X$ .Since $V \subset X \subset L^2(\Omega)$ and using the Poincaré–Friedrichs inequality, the spaces and norms are nested as

$$V \subset X \subset L^2(\Omega) \text{ and } ||f||_* \leq ||f||_{-1} \leq \frac{1}{C_{PF}}||f||.$$

We don't want to go too far afield into the question of existence of solutions to the Stokes problem in various Sobolev spaces with data in other Sobolev spaces. However, there is one very important theoretical result (proved by Ladyzhenskaya), the continuous inf-sup condition, which we give here for future reference.

[27]Of course, many norms are dual norms with respect to some other function space, so calling $||\cdot||_*$ *the* dual norm is imprecise, although commonly done in finite element CFD.

Proposition 5 (the continuous inf-sup condition). *There is a constant $\beta > 0$ such that*

$$\inf_{q\in Q}\sup_{v\in X}\frac{(q,\nabla\cdot v)}{||\nabla v||\,||q||}\geq\beta>0. \tag{4.2.1}$$

To indicate the critical importance of the last lemma and proposition, note that the lemma implies V is in fact also a Hilbert space. Since $\nabla\cdot u=0$ (weakly) the solution of the Stokes problem lies in V. It thus has the following formulation in V:

$$\text{find } u\in V \text{ satisfying } (\nabla u,\nabla v)=(f,v)\ \forall\, v\in V. \tag{4.2.2}$$

In this formulation, existence and uniqueness of u follow from the Lax–Milgram theorem, as does the simple bound on the fluid velocity:

$$||\nabla u||\leq||f||_*,\ \text{where } ||f||_*:=\sup_{v\in V}\frac{(f,v)}{||\nabla v||}.$$

This bound shows that the velocity is bounded by the body force, the most fundamental of many different types of stability. Thus, Lemma 5 and the Lax–Milgram theorem immediately imply existence of a velocity satisfying the variational formulation of the Stokes problem in V. (The inf-sup condition plays the key role of ensuring that, given the unique velocity, there is a corresponding pressure.)

Proposition 6 (existence). *For any $f\in L^2(\Omega)$, there exists a unique velocity in V solving the Stokes problem. This velocity u satisfies the á priori bound*

$$||\nabla u||\leq||f||_*.$$

The inf-sup condition is also critical to bounding the fluid pressure, i.e., showing the pressure is stable in a fundamental sense. To see this, note that the inf-sup condition (4.2.1) is equivalent to

$$\beta||q||\leq\sup_{0\neq v\in X}\frac{(q,\nabla\cdot v)}{||\nabla v||}\ \text{for any } q\in Q.$$

To use this, our strategy is to isolate the $(p,\nabla\cdot v)$ term and seek an upper bound resembling

$$(p,\nabla\cdot v)=\text{everything else}\leq\cdots\leq\{\text{terms}\}\ ||\nabla v||.$$

Next, we divide both sides by $||\nabla v||$ and take a supremum over $v\in X$. To be specific, rearranging (1.2),

$$\begin{aligned}(p,\nabla\cdot v)&=(\nabla u,\nabla v)-(f,v)\leq\\&\leq||\nabla u||\,||\nabla v||+||f||_*||\nabla v||\\&\leq(||\nabla u||+||f||_*)||\nabla v||.\end{aligned}$$

Thus,

$$\beta||p||\leq\sup_{0\neq v\in X}\frac{(p,\nabla\cdot v)}{||\nabla v||}\leq\{||\nabla u||+||f||_*\}.$$

The upper bound $||\nabla u|| \leq ||f||_*$ gives

$$||p|| \leq \beta^{-1} \cdot 2||f||_*.$$

Thus, adding the bounds for the velocity and pressure together gives

$$||\nabla u|| + ||p|| \leq (1 + \beta^{-1}2)||f||_*,$$

proving a stability bound on the fluid velocity and pressure.

4.3 The Galerkin Approximation

The finite element approximation of the Stokes problem begins by choosing finite dimensional spaces $X^h \subset X$ and $Q^h \subset Q$ and calculate $(u^h, p^h) \in (X^h, Q^h)$ as follows:

$$\begin{aligned} &\text{find } (u^h, p^h) \in (X^h, Q^h) \text{ satisfying} \\ (\nabla u^h, \nabla v^h) - &(p^h, \nabla \cdot v^h) = (f, v^h) \text{ and} \\ &(q^h, \nabla \cdot u^h) = 0 \, \forall \, (v^h, q^h) \in (X^h, Q^h). \end{aligned} \tag{4.3.1}$$

Without further details about (X^h, Q^h) it is not even true that (u^h, p^h) can be expected to exist. For example, the most natural choice of velocity-pressure finite element spaces is $X^h = C^0$ piecewise linears and $Q^h =$ piecewise constants. This choice fails! The following example, illustrated in Figure 4.4, of the linear-constant element pair on a very simple domain (Ω a hexagon in the plane) decomposed into six triangles illustrates this very well. (This nice example was constructed in Girault and Raviart [44].) Pick the pressure space to be piecewise constants on the given triangulation and the velocity space to be conforming piecewise linears for each component of the velocity. With discontinuous *constant* pressures the discrete incompressibility condition $(q^h, \nabla \cdot u^h) = 0$ becomes the five[28] local conditions

$$\int_\Delta \nabla \cdot u^h dx = 0 \text{ for all triangles } \Delta \text{ in the mesh.}^{29}$$

The pressure space has one degree of freedom (one at the centroid of each triangle) per triangle. The velocity space has two degrees of freedom (one per velocity component) per vertex. Counting degrees of freedom and constraints gives

$$\begin{aligned} X^h \; &: \; 7 \text{ nodes and 1 unknown per component per node} = 14 \text{ unknowns}, \\ u_j^h &= 0 \text{ at boundary nodes for } j = 1, 2\text{: 12 equations.} \\ \int_\Delta u^h dx &= 0 \, \forall \, \Delta \text{ in mesh: 5 linearly independent equations.} \end{aligned}$$

Thus we have already 17 linearly independent, homogeneous equations for only 14 unknowns. The condition that $u^h = 0$ on $\partial\Omega$ and $(\nabla \cdot u^h, q^h) = 0 \, \forall \, q^h \in Q^h$ produces

[28]Why five and not six? This is because the pressure space must be constrained to have mean value zero over Ω. This reduces the number of linearly independent equations by 1.

[29]This is equivalent to inflow = outflow on each triangle.

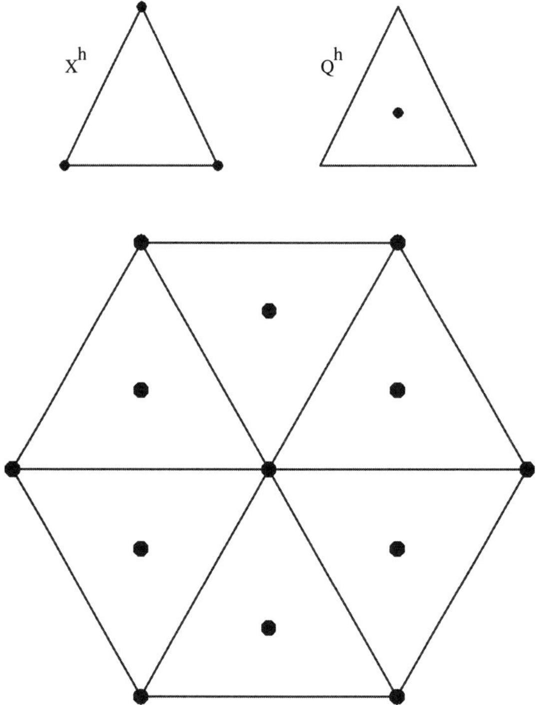

Figure 4.4. *Linear-constant pair violates stability.*

an overdetermined system for u^h's degrees of freedom, implying already $u^h \equiv 0$. This contradicts the nonzero information the momentum equation contributes to u^h!

This example demonstrates that the usual and most natural FEM for the Stokes problem can fail dramatically. Usually the failure mode is less dramatic: nonphysical oscillations in the pressure, often called the checkerboard mode, are observed. The Galerkin approximation in (X^h, Q^h) is thus not so clear and simple. Let's begin with a little lemma to try to simplify it!

Lemma 6 (the little lemma). *Let Y be a Hilbert space and $P_h, 0 < h \leq 1$, a family of projection operators with ranges the subspaces $Y^h \subset Y$:*

$$P_h : Y \to Y^h,\ P_h y^h = y^h\ \forall\ y^h \in Y^h.$$

Suppose Y^h is finite dimensional and P_h has uniformly bounded operator norm:[30]

$$\sup_{0<h\leq 1} ||P_h||_{L(Y,Y)} \leq \alpha < \infty.$$

[30]In other words, for all $y \in Y$,

$$||P_h y|| \leq \alpha ||y||.$$

Then, for any $w \in Y$

$$||w - P_h w||_Y \leq C(\alpha) \inf_{v^h \in Y^h} ||w - v^h||_Y.$$

We wish to apply this little lemma to the Galerkin approximation of the Stokes problem. Accordingly, let

$$Y := (X, Q),\ Y^h := (X^h, Q^h).$$

We want to define $P_h : Y \to Y^h$ by

$$P_h(u, p) = (u^h, p^h).$$

Naturally, we must show (u^h, p^h) exists and P_h is a well-defined projection operator!

Lemma 7. *Let* (u, p) *be a solution to the Stokes problem. If any solution* (u^h, p^h) *to* (3.1) *satisfies the á priori bound*

$$||(u^h, p^h)||_Y \leq C||(u, p)||_Y, \tag{4.3.2}$$

then (u^h, p^h) *exists and* $P_h(u, p) = (u^h, p^h)$ *is a well-defined projection operator.*

Proof. It's helpful in the proof to use the form of Galerkin orthogonality that (u^h, p^h) satisfies

$$\begin{aligned} &(\nabla(u - u^h), \nabla v^h) - ((p - p^h), \nabla \cdot v^h) = 0, \\ &(\nabla \cdot (u - u^h), q^h) = 0\ \forall\ (v^h, q^h) \in Y^h. \end{aligned} \tag{4.3.3}$$

Introducing a basis for (X^h, Q^h), (4.3.1) reduces to a linear system for the undetermined coefficients. Existence is thus implied by uniqueness, which is implied by its null-space being trivial. Set thus $u = 0$, $p = 0$ in (4.3.3). The á priori bound (4.3.2) implies immediately $(u^h, p^h) = 0$. To show P_h is a projection operator, i.e., $P_h^2 = P_h$, suppose $(u, p) \in Y_h$. The above argument also immediately implies $u^h = u$, $p^h = p$, i.e., $P_h^2 = P_h$. □

This lemma shows that the key to analyzing the Galerkin approximation is to prove it is bounded in the sense of (4.3.2). To isolate the main ideas, let us try to prove such a bound by direct assault.[31] The *Galerkin orthogonality* equations (4.3.3) can be rearranged as

$$(\nabla u^h, \nabla v^h) = (\nabla u, \nabla v^h) - (p - p^h, \nabla \cdot v^h)$$

and since $\nabla \cdot u = 0$, $(\nabla \cdot u^h, q^h) = 0$. Set $v^h = u^h \in X^h$ and $q^h = p^h \in Q^h$. This gives

$$||\nabla u^h||^2 = (\nabla u, \nabla u^h) - (p, \nabla \cdot u^h) + \underbrace{(p^h, \nabla \cdot u^h)}_{zero}.$$

Thus,

$$||\nabla u^h||^2 \leq ||\nabla u||\ ||\nabla u^h|| + ||p||\ ||\nabla \cdot u^h||.$$

[31]This means put the terms that give us the norm we want on the left-hand side and the rest on the right-hand side. The Cauchy–Schwarz inequality is applied to the right-hand side and "bad" terms are hidden in the "good" ones on the left-hand side.

As $||\nabla\cdot u^h|| \leq \sqrt{d}||\nabla u^h||$ we get the first half of the stability bound on the approximate velocity:

$$||\nabla u^h|| \leq ||\nabla u|| + \sqrt{d}||p||.$$

This stability bound exactly parallels the one in the previous section on the continuous velocity. The proof of the stability bound for the continuous pressure suggests that the following relation between X^h and Q^h is needed for boundedness of the discrete pressure. This is known as the *discrete inf-sup condition*, the *condition for div-stability*, and as the *(LBB^h) condition.*

Condition 1 (the Ladyzhenskaya–Babuska–Brezzi condition LBB^h). *Suppose* (X^h, Q^h) *satisfies*

$$\inf_{q^h\in Q^h} \sup_{v^h\in X^h} \frac{(q^h, \nabla\cdot v^h)}{||\nabla v^h||\,||q^h||} \geq \beta^h > 0, \tag{4.3.4}$$

where β^h *is bounded away from zero uniformly in* h.

The (LBB^h) condition is equivalent to

$$\beta^h||q^h|| \leq \sup_{v^h\in X^h} \frac{(q^h, \nabla\cdot v^h)}{||\nabla v^h||}.$$

Using this last form, it's easy to mimic the continuous argument and prove the following.

Lemma 8 (stability of velocity and pressure). *If* (LBB^h) *holds,*

$$||\nabla u^h||^2 + ||p^h||^2 \leq C(\beta^h)(||\nabla u||^2 + ||p||^2).$$

As a direct result of this and the little lemma we began with, we can conclude convergence of the FEM for the Stokes problem.

Theorem 13 (convergence). *Suppose* (LBB^h) *holds. Then, the Stokes projection* P_h *, given by*

$$P_h(u, p) = (u^h, p^h),$$

is well defined and its norm is bounded uniformly by $C(\beta^h)$. *Thus,*

$$||\nabla(u - u^h)||^2 + ||p - p^h|| \leq C(\beta^h) \inf_{\substack{v^h\in X^h\\ q^h\in Q^h}} \{||\nabla(u - v^h)||^2 + ||p - q^h||^2\}.$$

Proof. The proof follows immediately from the little lemma and the stability bound in the previous lemma. □

Since $(v^h, q^h) \to b(v^h, q^h)$ is continuous, the space of discretely divergence-free functions, V^h, defined by

$$V^h = \{v^h \in X^h : (q^h, \nabla\cdot v^h) = 0\ \forall\, q^h \in Q^h\},$$

is a closed subspace of X^h. Further, under (LBB^h) the formulations of the discrete Stokes problem in V^h and in (X^h, Q^h) are equivalent. The condition (LBB^h) also implies that V^h is nontrivial in the sense that V^h has an interesting and important approximation property.

Lemma 9. *Let $u \in V$ and suppose (LBB^h) holds. Then,*

$$\inf_{v^h \in V^h} ||\nabla(u - v^h)|| \le C(\beta^h) \inf_{v^h \in X^h} ||\nabla(u - v^h)||.$$

Proof. This is proven in Girault and Raviart [44]. □

4.4 More About the Discrete Inf-Sup Condition

The simplest examples of linear velocities and constant pressures of Girault and Raviart [44] clearly violate the condition (LBB^h). Fortunately, there are many examples of "good" velocity-pressure spaces which do satisfy (LBB^h). These are catalogued in books such as Gunzburger [46], Girault and Raviart [45], and Brezzi and Fortin [16] . To understand this essential condition we shall show how one specific element is constructed to satisfy (LBB^h): the MINI element of Arnold, Brezzi, and Fortin [6].

To motivate it, reconsider (LBB^h):

$$\inf_{q^h \in Q^h} \sup_{v^h \in X^h} \frac{(q^h, \nabla \cdot v^h)}{||\nabla v^h|| \, ||q^h||} \ge \beta^h > 0. \tag{4.4.1}$$

Lemma 10. *Suppose (LBB^h) holds in (X^h, Q^h). Then it also holds in*

(a) $(X^h, \tilde{Q}^h)$ *if* $\tilde{Q}^h \subset Q^h$,

(b) $(\tilde{X}^h, Q^h)$ *if* $\tilde{X}^h \supset X^h$.

Proof. Taking the supremum over a larger set increases its value, and taking the infimum over a smaller set decreases the infimum. □

This lemma suggests that if (X^h, Q^h) doesn't quite satisfy (LBB^h) we should try to modify it by making X^h larger or Q^h smaller!

There are various methods for verifying the inf-sup condition in finite element spaces. One good approach is originally due to Fortin. It provides conditions under which (4.4.1) can be reduced to (or inherited from) the continuous inf-sup condition.

Lemma 11 (Fortin's lemma). *Let $X^h \subset X$, $Q^h \subset Q$. Suppose there is a linear operator*

$$\mathcal{F}^h : X \to X^h$$

such that $||\mathcal{F}^h||$ is bounded, uniformly in h, i.e.,

$$||\nabla(\mathcal{F}^h v)|| \le C||\nabla v|| \quad \forall\, v \in X,$$

and also such that for all $v \in X$

$$(\nabla \cdot (v - \mathcal{F}^h v), q^h) = 0 \quad \forall\, q^h \in Q.$$

Then, (4.4.1) *, i.e., condition* (LBB^h)*, holds.*

Proof. The following chain of inequalities holds:

$$
\begin{aligned}
&\inf_{q^h\in Q^h}\sup_{v^h\in X^h}\frac{(q^h,\nabla\cdot v^h)}{||\nabla v^h||\,||q^h||}\\
&\geq \inf_{q^h\in Q^h}\sup_{v\in X}\frac{(q^h,\nabla\cdot(\mathcal{F}^h v))}{||\nabla(\mathcal{F}^h v)||\,||q^h||}\\
&\geq \inf_{q^h\in Q^h}\sup_{v\in X}\frac{(q^h,\nabla\cdot v)}{||\mathcal{F}^h||\,||\nabla v||\,||q^h||}\\
&\geq \frac{1}{||\mathcal{F}^h||}\inf_{q\in Q}\sup_{v\in X}\frac{(q,\nabla\cdot v)}{||\nabla v||\,||q||}\geq\frac{\beta}{||\mathcal{F}^h||}. \qquad \square
\end{aligned}
$$

With these two tools in hand, we can now present the MINI element and give the essential idea behind the proof that (LBB^h) holds. Consider a typical element $\triangle$ in a finite element triangulation of $\Omega\subset\mathbb{R}^2$. Let $\lambda_1\lambda_2\lambda_3(=\lambda_j(x,y))$ denote the barycentric coordinates of $\triangle$. These are just the standard, piecewise linear basis functions on $\triangle$ for which $\lambda_j=1$ at the jth vertex and λ_j vanishes on the edge opposite this vertex. Consider the cubic that is the product of these three local basis functions,

$$\phi_\triangle(x,y)=\lambda_1(x,y)\,\lambda_2(x,y)\,\lambda_3(x,y).$$

$\phi_\triangle$ is a cubic polynomial on $\triangle$. It vanishes on $\partial\triangle$ (and also outside of $\triangle$). This special cubic is called the cubic bubble function due to its shape. The MINI element consists of C^0 piecewise linears enhanced by cubic bubble functions for the velocity and C^0 piecewise linears for the pressure.

$$
\begin{aligned}
X^h &= \{v^h\in H_0^1(\Omega): v^h|_\triangle\in\mathcal{P}_1(\triangle)\ \forall\triangle \text{ in the mesh}\}\oplus\{\phi_\triangle:\forall\triangle\text{ in the mesh}\},\\
Q^h &= \{q^h\in L_0^2(\Omega)\cap C^0(\Omega): q^h|_\triangle\in\mathcal{P}_1(\triangle)\ \forall\triangle\text{ in the mesh}\}.
\end{aligned}
$$

It is represented in Figure 4.5.

Theorem 14. *The MINI element satisfies the* (LBB^h) *condition.*

We will soon give a sketch of the essential ideas of the proof. First, note that the following is a direct consequence.

Proposition 7. *Let* (u^h,p^h) *be the finite element approximation of the Stokes problem with the MINI element. Then,*

$$||\nabla(u-u^h)||+||p-p^h||\leq Ch(||\nabla\nabla u||+||\nabla p||).$$

Proof **(sketch of the proof of Theorem 14).** We shall construct an operator $\mathcal{F}^h: X\to X^h$ satisfying the conditions of Fortin's lemma. Calling $v^h=\mathcal{F}^h v$, we seek to construct v^h (given v) with

$$0=\int_\Omega q^h\nabla\cdot(v-v^h)dx\ \ \forall q^h\in Q^h.$$

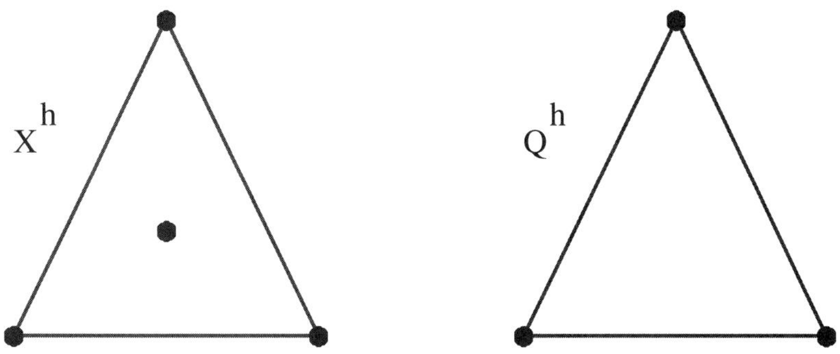

Figure 4.5. *The MINI element.*

Since Q^h is continuous, we can integrate this by parts and obtain that this is equivalent to

$$0 = \int_\Omega (\nabla q^h) \cdot (v - v^h) dx \ \ \forall\, q^h \in Q^h.$$

It clearly suffices for this to hold on every triangle individually, or

$$\int_\Delta (\nabla q^h) \cdot (v - v^h) dx = 0 \quad \forall\, \Delta \text{ in the mesh.}$$

On a fixed (but arbitrary) triangle Δ, $q^h|_\Delta = a + bx + cy$ so $\nabla q^h|_\Delta = (b, c)^{tr}$.
Thus, sufficient for (LBB^h) is

$$\text{(key condition)} \int_\Delta v_j - v_j^h dx = 0, \ \ j = 1, 2, \ \ \forall\, \Delta \text{ in the mesh.}$$

Now X^h is the span of the (vertex) nodal piecewise linear basis functions λ_j and the bubble functions. Defining v_j^h is equivalent to specifying v_j^h at each node.

Recall that the support of a function ϕ, supp(ϕ), is closure$\{x : \phi(x) \neq 0\}$. At each vertex node, N_k inside Ω define

$$(\text{for } j = 1, 2)\ v_j^h(N_k) = \int_{support\ (\phi_k)} v_j(x) dx / meas\ (supp\ \phi_k).$$

Thus

$$v_j^h = \sum_{\text{all vertices } N_k \text{ inside } \Omega} v_j^h(N_k)\phi_k(x) + \sum_{\text{all } \Delta} \alpha_{j\Delta}\phi_\Delta(x),$$

where ϕ_Δ is the cubic bubble function. The $\alpha_{j\Delta}$, $j = 1, 2$, are determined by requiring

$$\alpha_{j\Delta} \int_\Delta \phi_\Delta(x) dx = \int_\Delta v_j(x) dx - \int_\Delta \sum_{\text{all nodes } N_k \text{ inside } \Omega} v_j^h(N_k)\phi_k(x) dx.$$

Clearly, the (key condition) holds, by construction. That $||v^h||_X \leq C||v||_X$ follows by a scaling argument we shall omit. $\square$

4.4.1 Other div-stable elements

A choice of velocity-pressure finite element spaces which satisfy (LBB^h) is often called a *div-stable element.*[32] Many div-stable elements are known. For example, all the higher order cousins of the (unstable) linear-constant pair, given by

$$X^h = \{v^h \in H_0^1(\Omega) : v^h|_\Delta \in \mathcal{P}_k(\Delta) \ \forall \ \Delta \text{ in the mesh}\},$$
$$Q^h = \{q^h \in L_0^2(\Omega) \cap C^0(\Omega) : q^h|_\Delta \in \mathcal{P}_{k-1}(\Delta) \forall \Delta \text{ in the mesh}\}$$

are div stable. The choice $k = 2$ is known variously as the Taylor–Hood or Hood–Taylor element. It is a very popular choice in finite element CFD.

Theorem 15. *For any $k > 1$ the above choice of velocity pressure spaces X^h, Q^h satisfies* (LBB^h) .

Proof. See, for example, [16], [45]. □

Thus, most difficulties with div-stability occur when using low-order element finite element spaces. Unfortunately, these are also the ones most commonly used for difficult flow problems (for which the required resolution approaches the storage capacities of many computers).

One interesting low-order element that is div-stable is the nonconforming linear-constant Crouzeix–Raviart element [20]. Functions in this velocity space are not continuous across element edges, only approximately so in the sense that discontinuities have mean value zero across each edge in the mesh. Many nonconforming and quadrilateral div-stable elements are also known [16], [45].

4.5 Remarks on Chapter 4

The *pressure* in an incompressible flow is not a thermodynamic variable (for example, no gas law is determining it). It is often referred to as a Lagrange multiplier associated with the incompressibility constraint. This association is made rigorous by reformulating the Stokes problem as a constrained minimization problem, beautifully presented in [44].

The MINI element is a good element that satisfies condition LBB^h for stability of the pressure. Many other excellent elements exist, such as the Hood–Taylor element of quadratic velocities and linear pressures. Both the MINI element and the Hood–Taylor element use continuous pressure spaces. Since the discrete continuity equation is

$$\int_\Omega q^h \nabla \cdot v^h dx = 0 \ \forall q^h \in Q^h,$$

continuous pressures lead to a global discretizations of incompressibility. If Q^h consists of functions discontinuous across element edges, the incompressibility is localized to

$$\int_\Delta q^h \nabla \cdot v^h dx = 0 \ \forall \Delta \text{ in the mesh.}$$

[32]This is a common description but a bit odd at first sight since the inf-sup condition implies stability of the pressure.

There are flow problems for which this is preferable. This chapter has only scratched the surface on the theory of mixed FEMs. (In fact, many interesting and powerful aspects of the theory have been suppressed.) A second look at the theory should begin with either of the books by Girault and Raviart [45] or Gunzburger [46]. The MINI element paper of Arnold, Brezzi, and Fortin [6] is still a delight to read, and the current state of the art is well presented in Brezzi and Fortin [16].

Proposition 6 and the continuous inf-sup condition together imply existence of a pressure p such that u, p satisfies the variational formulation of the Stokes problem in (X, Q), i.e., that the formulations are all equivalent. Indeed, the problem for p now reads:

$$\text{given } v \in V \text{ find } p \in Q \text{ satisfying,}$$
$$(p, \nabla \cdot v) = (\nu \nabla u, \nabla v) - (f, v)\ \forall v \in X.$$

Since u exists, the right-hand side is a known linear functional,

$$L : X \to R, \text{ by } L(v) := (\nu \nabla u, \nabla v) - (f, v).$$

Thus, the problem of the pressure is to find $p \in Q$ satisfying

$$b(v, p) = L(v)\ \forall v \in X.$$

This is almost enough to prove existence of p using Babuska's generalization of the Lax–Milgram theorem and the inf-sup condition. There are several technical steps in completing this proof, however. Briefly, when the test function v is divergence free, $b(v, p) = 0$, so the above equation gives no information on the pressure. To obtain a variational equation for the pressure to which the Lax–Milgram theorem can be applied, the above is restricted to test functions in the X-orthogonal complement of V, $V^{\perp}$. At a key step, some characterization of functions in $V^{\perp}$ is needed to verify the positivity conditions needed. Thus, deeper results (like the existence of the pressure) also follow by combinations of the tools we are developing. The other related approach is to apply Babuska's generalization of the Lax–Milgram theorem. To verify the essential inf-sup condition, given a pressure p we need to specify a velocity v so that $b(v, p)$ is positive. By the above $v \in V^{\perp}$; by the Helmholtz theorem, such a v is curl free so choosing $v = \nabla \phi$ is a good try. Picking v with $\nabla \cdot v = p$ (and thereby verify the required positivity condition) reduces to solving the Poisson problem $\nabla \cdot \nabla \phi = p$. There are technical aspects to this approach as well.

Various vector identities and $\nabla \cdot u = 0$ can be used to derive several other possible formulations of the viscous term for the Stokes problem. These include

$$(\nabla u, \nabla v), 2(\nabla^s u, \nabla^s v), \text{ and } (\nabla \times u, \nabla \times v)$$

and weighted averages of the above. These are equivalent for the continuous problem with no-slip boundary conditions. For the discrete problem or with derivative boundary conditions they are not equivalent. Since $\nabla \cdot u = 0$ and $\nabla \cdot u^h \neq 0$ (in general) we can also add a term of the form $+\alpha(\nabla \cdot u^h, \nabla \cdot v^h)$ to the discrete Stokes problem.

A posteriori error estimates can also be developed for flow problems for velocity errors in various norms, for pressure errors, and even for errors in local averages in flow variables, e.g., [59] and Hoffman [52]. Generally, as problems become more complex, adaptive meshes become more important and, as a consequence, so do a posteriori error estimates.

4.6 Exercises

Exercise 36 (the pressure Poisson equation). *If u, p satisfies the Stokes problem and is smooth enough, show (by taking the divergence of the Stokes problem and rearranging) that p satisfies the following pressure Poisson equation forced by both $f(x)$ and the velocity:*

$$-\Delta p = -\nabla \cdot f \textit{ in } \Omega \textit{ and}$$
$$\nabla p \cdot n = (f + \Delta u) \cdot n \textit{ on } \partial\Omega.$$

Exercise 37. *Show that if (u, p) satisfies* (4.1.1), *then u satisfies* (4.2.2).

Exercise 38. *Complete the indicated proof that $u \in V$ exists.*

Exercise 39 (uniqueness of the pressure). *Prove that p is unique directly using the inf-sup condition.*

Exercise 40 (stability). *Prove $||(u, p)||_Y \le C(\beta)||f||_*$. (**Hint:** Use the definition of the dual norm of f.)*

Exercise 41 (porous media problem). *Consider the porous media problem for*

$$0 < k_{\min} \le k(x, y) \le k_{\max} < \infty.$$

Give its mixed variational formulation. Formulate an inf-sup condition and prove

$$||(u, \phi)||_Y \le C||f||_*.$$

Exercise 42 (existence of the pressure). *Do a literature search and find a detailed and complete proof that p exists. Rewrite the proof into one using the generalization of the Lax–Milgram theorem.*

Exercise 43. *Prove Lemma* 9. *If you can't, find and write out a detailed explanation of the proof in Girault and Raviart [44].*

Exercise 44 (stability of the discrete pressure). *Prove the lemma asserting stability of the pressure under LBB^h.*

Exercise 45 (the little lemma). *Prove the little lemma. (**Hint:** First show that $P_h = I$ on Y^h implies $Rge(P_h) = Y^h$. Next, use the idea of proof of Céa's lemma where the error is decomposed into a part inside Y^h and a part outside Y^h.)*

Exercise 46 (another proof of Céa's lemma). *Use the little lemma to give an alternate proof of Céa's lemma. (**Hint:** Define $P_h u = u_h$.)*

Exercise 47. *Give the justification for each step in the chain of inequalities in the proof of Fortin's lemma.*

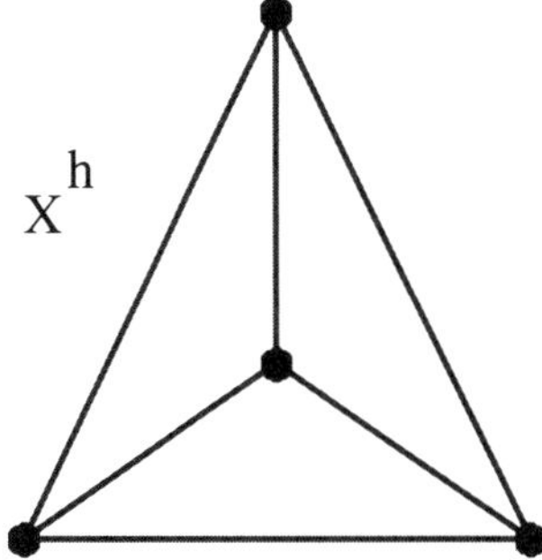

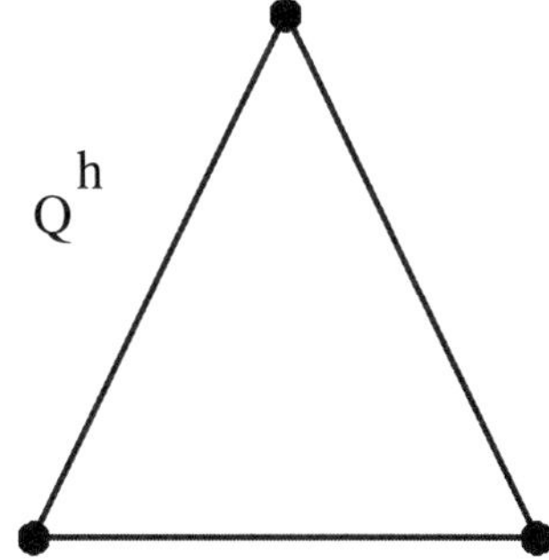

Figure 4.6. *Another element satisfying the discrete inf-sup condition.*

Exercise 48. *Let Q^h denote the space of C^0 piecewise linear functions with mean value zero over Ω. Take the mesh for Q^h and do one refinement as pictured in Figure 4.6. Define X^h to be finite element space of (vector valued) linears on the new mesh, vanishing on $\partial\Omega$.*

Show that (X^h, Q^h) satisfies the discrete inf-sup condition.

Exercise 49 (penalty methods). *There are many other formulations of the Stokes problem. For example, a penalty method replaces*

$$-\Delta u + \nabla p = f, \nabla \cdot u = 0$$

by (where $\epsilon = \epsilon(h)$ is small)

$$-\Delta u_\epsilon + \nabla p_\epsilon = f, \ \nabla \cdot u_\epsilon + \epsilon p_\epsilon = 0.$$

(a) *Find the variational formulation of (u_ϵ, p_ϵ) (note that p_ϵ can be eliminated).*

(b) *Formulate the penalty-finite element method and prove*

$$||\nabla u_\epsilon^h|| + ||p_\epsilon^h|| \leq C(\epsilon)(||\nabla u|| + ||p||).$$

Exercise 50 (coercivity versus the inf-sup condition). *The simplest example of coercivity versus an inf-sup condition is solving the 2×2 linear system $Au = f$ or*

$$\begin{bmatrix} a & b \\ c & d \end{bmatrix} \begin{bmatrix} u_1 \\ u_2 \end{bmatrix} \begin{bmatrix} f_1 \\ f_2 \end{bmatrix}.$$

(a) *Show that this is equivalent to: find $u \in \mathbb{R}^2$ such that*

$$a(u, v) := (v_1, v_2) \begin{pmatrix} a & b \\ c & d \end{pmatrix} \begin{pmatrix} u_1 \\ u_2 \end{pmatrix} = (v_1, v_2) \begin{pmatrix} f_1 \\ f_2 \end{pmatrix}$$

for all $v \in \mathbb{R}^2$.

(b) *Show that this has a solution for all* f *if and only if there is a* $C > 0$ *such that every solution* u *satisfies*

$$|u| \leq C|f|.$$

Show that in this case the above $a(\cdot, \cdot)$ *satisfies an inf-sup condition.*

(c) *Formulate coercivity for* $a(u, v)$ *and show that* $a(\cdot, \cdot)$ *is coercive if and only if the symmetric part of* A *is positive definite.*

Chapter 5
The Equations of Fluid Motion

Now I think hydrodynamics to be the root of all physical science, and is at present second to none in the beauty of its mathematics.

William Thompson (Lord Kelvin) (1824–1907), in a letter to Stokes, December 20, 1857, in *The Correspondence Between Sir George Gabriel Stokes and Sir William Thompson Baron Kelvin of Larges*, Cambridge University Press, Cambridge, 1990.

The extensive researches of V. Bjerknes and his school are pervaded by the idea of using differential equations for all they're worth.

L. F. Richardson (1881–1953), in *Weather Prediction by Numerical Process*, Dover, NY, 1965.

5.1 Conservation of Mass and Momentum

Although I envy a great generality with regard to the nature of fluids and the forces that are being applied to their particles, I have no fear of the reproaches often leveled with good reasons at those who have undertaken to generalize the researches of others. Often a great generality causes more confusion than it does illuminate, and sometimes it leads to such voluminous computations that it becomes extremely difficult to derive any consequences, even in the simplest cases. If generalizations have this disadvantage, then we should retreat from them and restrict ourselves to the special cases.

The generality that I embrace, far from dazzling our lights, will reveal to us rather the veritable laws of Nature in all their brilliance, and in them we shall find even stronger reasons to admire her beauty and her simplicity. It will be an important lesson to learn that some principles, till now believed to be bound to some special cases, are of greater breadth. Finally, these researches will demand calculations scarcely any more troublesome, and it will be easy to apply them to all special cases we might set up.

L. Euler, in *General Principles of the State of Equilibrium of Fluides*, 1755.

The Navier–Stokes equations (NSE) are a continuum model for the motion of a fluid. There are various ways to develop the NSE. For example, the Boltzmann equation describes the motion of molecules in a rarefied gas. NSE can follow by taking spatial averages of the Boltzmann equation. It can likewise arise from the kinetic theory of gasses. It has even been derived from quantum mechanics by a suitable averaging procedure.

The approach in this chapter is the more classical approach of continuum mechanics in which all the flow variables,

$$\text{density } \rho, \quad \text{velocity } u, \quad \text{pressure } p, \ldots,$$

are assumed to be continuous functions of space and time from the beginning. This approach can be made completely axiomatic. This chapter will try to find a middle path which is axiomatic enough but still retains a connection to the physical ideas.

Conservation of Mass

The equation describing conservation of mass is called the continuity equation. If mass is conserved, the rate of change of mass in a volume V must equal the net mass flux across ∂V:

$$\frac{d}{dt}\int_V \rho\, dx = -\int_{\partial V} (\rho u)\cdot \hat{u}\, d\sigma.$$

The divergence theorem thus implies

$$\int_V \frac{\partial \rho}{\partial t} + \nabla \cdot (\rho u) dx = 0.$$

If all the variables are continuous, shrinking V to a point gives

$$\frac{\partial \rho}{\partial t} + \nabla \cdot (\rho u) = 0,$$

which is the first equation of mathematical fluid dynamics. If the fluid is incompressible (and homogeneous),[33]

$$\rho(x,t) \equiv \rho_0,$$

and conservation of mass reduces to

$$\nabla \cdot u(x,t) = 0,$$

which is a constraint on the fluid velocity, u.

Loosely speaking, if a force tries to compress an incompressible fluid (think: a liquid), its density does not change but the pressure responds instead by increasing. For example, water is quite heavy, so its weight alone acts as a compressive force. However, the density of water at the bottom of the ocean is very close to its density at the surface, but the pressure at the ocean's floor is much greater than at the surface.

[33] Homogeneous means composed of one fluid. It is possible to have an incompressible mixture of two incompressible fluids (liquids) in which the density changes because of variations in the fraction of each in the mixture. Fluid mixture models are important in CFD and built on a clear understanding of the equations for the motion of one fluid.

Conservation of Momentum

Conservation of momentum states that the rate of change of linear momentum must equal the net forces acting on a collection of fluid particles, or

$$\text{force} = \text{mass} \times \text{acceleration}.$$

Let us consider a fluid particle. If it is at (x, t) (position x at time t), then at time $t + \Delta t$ it has flowed to (up to the accuracy of the linear approximation)

$$(x + u\Delta t, t + \Delta t).$$

Its acceleration is therefore, by adding and subtracting terms in the limit,

$$\lim_{\Delta t \to 0} \frac{u(x + u(x,t)\Delta t, t + \Delta t) - u(x,t)}{\Delta t} = \frac{\partial u}{\partial t} + \sum_j u_j \frac{\partial u_i}{\partial x_j} = u_t + (u \cdot \nabla)u.$$

Thus, the mass $\times$ acceleration in a volume V is

$$\int_V \rho(u_t + u \cdot \nabla u)\, dx$$

and must be balanced by external (body) forces and internal forces acting on that volume.

External forces include gravity, buoyancy, and electromagnetic forces (in liquid metals). These are collected in a body force term which has accumulated net force on the volume V given by

$$\int_V f\, dx.$$

Internal forces are the forces that a fluid exerts on itself in trying to get out of its own way. These include pressure and the viscous drag that a fluid element exerts on the adjacent fluid. The internal forces of a fluid are *contact* forces: they act on the surface of the fluid element V.[34] If $\vec{t}$ denotes this internal force vector, called the *Cauchy stress vector* or *traction vector*, then the net contribution of the internal forces on V is

$$\int_{\partial V} \vec{t}\, d\sigma$$

(where we use the arrow to distinguish between t = time and $\vec{t}$ = traction vector). Thus, the equation for conservation of momentum is, for all volumes V,

$$\int_V \rho(u_t + u \cdot \nabla u) dx = \int_V f dx + \int_{\partial V} \vec{t}\, d\sigma.$$

To derive a local, pointwise equation (a partial differential equation, for example) from an integral balance equation, the general plan is to divide by the volume of V, and then

[34]On the one hand, this is a fundamental physical fact about internal forces in a fluid. On the other hand, it is a fundamental postulate, first formulated by A.-L. Cauchy, in developing the continuum mechanics of fluids. There are various equivalent ways to restate the postulate. One way is, *The forces exerted on a fluid volume by the rest of the fluid can be represented as a distribution of stress vectors on the boundary of the fluid volume.*

shrink V to a point. For this plan to succeed, the last integral , $\int_{\partial V} \vec{t}\, d\sigma$, over ∂V must be replaced by a volume integral over V. To do so, we will need more information on the internal, contact forces represented by $\vec{t}$.

Modeling these internal forces correctly is critical to predicting the fluid motion correctly. We will look at these internal forces more carefully next.

Remark 4. *In the material derivative* $(u \cdot \nabla)u$ *is the vector*

$$((u \cdot \nabla)u)_j = u_i \frac{\partial}{\partial x_i} u_j = \sum_{i=1}^{d} u_i u_{j,x_i} = u_i u_{j,i}.$$

Since vector-tensor algebra is defined so as to be associative, it is equal to $u \cdot (\nabla u)$. *However, it is more useful to think of* $(u \cdot \nabla)u$ *as the operator* $\phi \to (u \cdot \nabla)\phi$ *acting on* $\phi = u$.

5.2 Stress and Strain in a Newtonian Fluid

I must confess that I do not know how the resistance of fluids can be explained theoretically, because the theory gives zero resistance....

d'Alembert, 1768, in "Paradoxe proposé aux géomètres sur résistance des fluides," *Opuscules mathématiques*, Vol. 5, 34th Memoir, Paris.

Fluids are subject to outside or external forces, such as gravity and forces acting at the boundary of a region (injection and suction come to mind). They are also subject to internal forces such as buoyancy. Fluids also exert forces which act intrinsically on the fluid motion as sections of the fluid exert drag on each other as they try to get out of each others way. The way these internal forces in a fluid act are the key to "fluidity"[35] and to the difference between solids, liquids, and gases. They also differentiate among fluids. The three laws of Cauchy are the foundation of fluid (and even continuum) mechanics. These laws give a precise description of the internal forces in a fluid based on three fundamental physical laws: internal forces are surface forces, conservation of linear momentum, and conservation of angular momentum.

The mathematical statement of the stress principle of Cauchy[36] is that *on any (imaginary) plane there is a net force that depends (geometrically) only on the orientation of that plane.* In other words, all (internal) forces exerted on a fluid volume by the fluid outside this volume can be accounted for by a distribution of vectors on the surface of the volume. This vector is called the traction or Cauchy stress vector, $\vec{t}$. It depends only on position, time, and the local orientation of the surface, i.e., on the normal , $\vec{t} = \vec{t}(x, t, \widehat{n})$. Suppressing the dependence on x, t we write

$$\vec{t} = \vec{t}(\widehat{n}), \widehat{n} = \text{normal vector to tangent plane.} \tag{5.2.1}$$

The exact dependence of $\vec{t}$ upon $\widehat{n}$ was determined rigorously by Cauchy (now often called Cauchy's theorem) using other (now accepted) principles of continuum mechanics. Cauchy

[35]The basic property which differentiates a fluid from a solid is that a fluid at rest cannot support tangential stresses. In other words, if you look at a surface in the fluid that is at rest, then the resultant of the oblique stresses to the surface, taken as one side limits from each side, must be zero.

[36]Actually, Euler had earlier given a very similar stress principle for the case of perfect fluids.

proved that if linear momentum is conserved, then

$$\vec{t} \text{ is a linear function of the normal vector } \widehat{n}.$$

Thus, the action of the Cauchy stress vector is determined by a 3×3 matrix, called the *stress tensor* $\Pi = \Pi(x, t)$ with

$$\vec{t}(\widehat{n}) = \widehat{n} \cdot \Pi.$$

The stress tensor represents all the internal forces in the flow of a fluid. With this stress tensor Π we can write the equation for conservation of linear momentum as follows.[37] For a (spatially) fixed volume V, applying the divergence theorem gives

$$\begin{aligned}\int_V \rho\left(\frac{\partial u}{\partial t} + u \cdot \nabla u\right) dx &= \int_{\partial V} \hat{t}(\hat{n}) d\sigma + \int_V f dx \\ &= \int_{\partial V} \widehat{n} \cdot \Pi d\sigma + \int_V f dx = \int_V \nabla \cdot \Pi + f dx.\end{aligned}$$

Shrinking V to a point then gives

$$\rho\left(\frac{\partial u}{\partial t} + (u \cdot \nabla) u\right) = \nabla \cdot \Pi + f \text{ in } \Omega,$$

which is called the *momentum equation* because it expresses $f = ma$ or conservation of linear momentum.

Conservation of angular momentum is also an important physical principle that any reasonable model of a physical system must satisfy. Cauchy's second law is that conservation of angular momentum requires that Π be symmetric:[38]

$$\Pi_{ij} = \Pi_{ji} \forall\, i, j.$$

In summary, the three important relations which follow from the Cauchy stress principle are

- $\vec{t}(\widehat{n}) = \widehat{n} \cdot \Pi$,
- $\Pi_{ij} = \Pi_{ji}$, and
- $\rho(\frac{\partial u}{\partial t} + (u \cdot \nabla) u) = \nabla \cdot \Pi + f$.

5.2.1 More about internal forces

A fluid has several types of internal forces:

- **Pressure forces = normal forces**

These act on a surface purely normal to that surface. The total, net normal force is the averaged of the diagonal of the stress tensor. We define the pressure in an incompressible

[37] Recall that the divergence of a tensor T is defined by $(\text{div } T)_i = T_{ij,j}$.

[38] Cauchy discovered the law and proved $\Rightarrow$ while Boltzmann completed the connection proving $\Leftarrow$.

flow (called the dynamic pressure) then as $p := \frac{1}{3}(\Pi_{11} + \Pi_{22} + \Pi_{33})$. The force exerted by the pressure is then

$$\text{pressure force} = -p\mathbf{I}\hat{n}, \text{ where } p(x,t) = \text{ the dynamic pressure.}$$

• **Tangential forces = Viscous forces**

The nonpressure part of the stress tensor, called the viscous stress tensor **V**, is defined by

$$\mathbf{V} := \Pi + p\mathbf{I}, \text{ so } \Pi = -p\mathbf{I} + \mathbf{V}.$$

One basic fact about fluids is that a fluid in physical *equilibrium*[39] cannot support tangential stresses. Thus, only pressure forces (or hydrostatic forces) can exist for a fluid at rest. (The first known book on mathematical fluid dynamics was *Hydrostatics* by Archimedes on precisely this subject.) Thus, differences between fluids[40] in motion must reside in the nonpressure part of the stress tensor, $\mathbf{V} = \Pi + p\mathbf{I}$.

Example 13 (the simplest fluid model is a *perfect fluid*). A perfect fluid is incompressible and without internal viscous forces; thus, $\mathbf{V} = 0$. If $\mathbf{V} = 0$, then the system is closed: it has the same number of equations as unknowns. The motion of a perfect fluid is governed by the Euler equations:

$$\nabla \cdot u = 0, \text{ and } u_t + u \cdot \nabla u + \nabla\left(\frac{p}{\rho_0}\right) = \frac{1}{\rho_0} f. \quad \blacksquare \tag{5.2.2}$$

In the general case, including nonzero viscous forces, the system of equations is not closed until **V** is related to the fluid velocity. Internal forces must depend on local velocity differences (rather than the velocity itself), so **V** should depend on some combination of derivatives of u. This combination is the deformation tensor **D**. The simplest relation is a linear law between stress and strain (force and deformation). This is the exact analogue for fluids of Hooke's law for solids. In fact, it makes more sense for fluids than for solids since if stress concentrates in a solid it breaks, while a fluid, in contrast, will simply move or flow out of its own way to decrease stress.

Assumption: Linear stress-deformation relation. *Let* $\mathbf{D} = \frac{1}{2}(\nabla u + \nabla u^t)$. *Assume that*

$$\mathbf{V} = 2\mu\,\mathbf{D} + \left(\xi - \frac{2\mu}{3}\right)\mathbf{I}\,\nabla \cdot u.$$

The parameters μ and ξ are the *first and second viscosities* of the fluid. The physical parameter μ is called the *dynamic or shear viscosity*.

5.2.2 More about *V*

It is good to keep in mind that a linear stress-strain relation is only an *approximation* for a real fluid. Schematically, the stress and deformation are related nonlinearly (in general and especially for large deformations); see Figure 5.1. For small stresses, a linear approximation near zero to the general stress-deformation relation can be used.

[39]"Physical equilibrium" means at rest whereas mathematical equilibrium means only time-independent.

[40]For nonisothermal flows of compressible fluids (gasses), thermodynamics add more possibilities for differences.

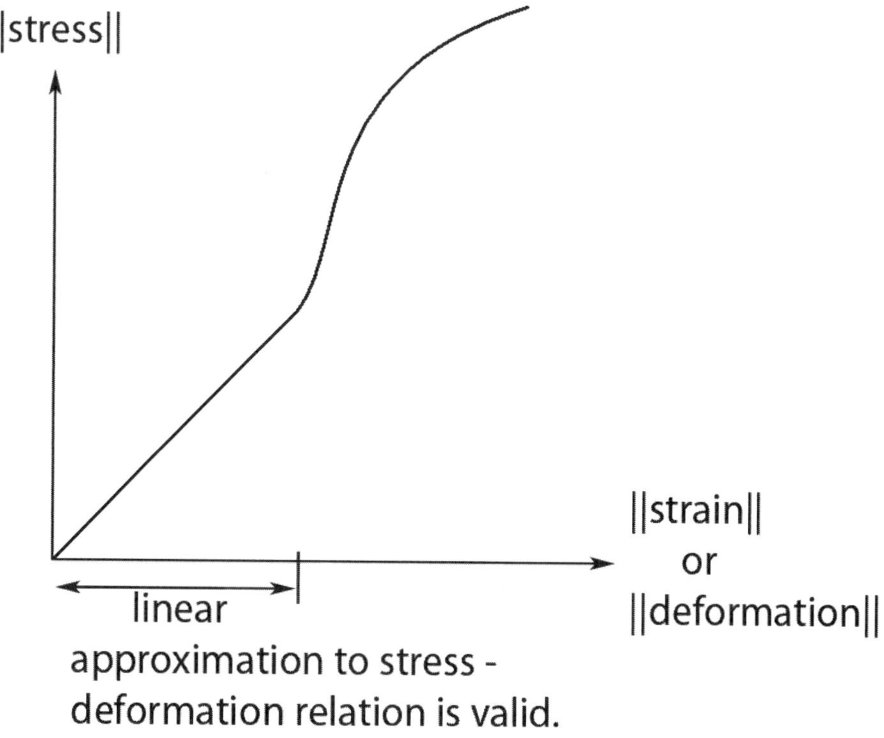

Figure 5.1. *Stress-deformation relation is nonlinear.*

The first scientist to postulate a linear stress-strain relation was Newton:

> *The resistance arising from the want of lubricity on the parts of the fluid is, other things being equal, proportional to the velocity with which the parts are being separated from one another.*
>
> I. Newton, *Principia*, Book 2, IX.

For this reason, a fluid satisfying this assumption is called a *Newtonian fluid.* More general relations for $\mathbf{V} = \mathbf{V}(\mathbf{D})$ exist and are appropriate for fluids with larger stresses. We shall not go into detail about these fluids herein other than to state that they are of great practical importance and are not well understood.

The NSE describe a fluid satisfying a linear stress-deformation relation. They are thus

$$\rho_t + \nabla \cdot (\rho u) = 0,$$
$$\rho\left(\frac{\partial u}{\partial t} + u \cdot \nabla u\right) + \nabla p - \nabla \cdot \left[2\mu\mathbf{D}(u) + \left(\xi - \frac{2\mu}{3}\right)\mathbf{I}\nabla \cdot u\right] = f,$$

holding in the flow domain Ω and over the time interval $0 < t \leq T$. If the fluid is incompressible and μ is constant, the NSE reduce to

$$\nabla \cdot u = 0 \text{ and } u_t + u \cdot \nabla u + \nabla \left(\frac{p}{\rho_0} \right) - \frac{\mu}{\rho_0} \Delta u = \frac{1}{\rho_0} f \text{ in } \Omega. \tag{5.2.3}$$

The pressure p is now simply redefined to be p/ρ_0. The parameter μ is the fluids' viscosity coefficient and

$$\mu/\rho_0 =: \nu = \text{ the kinematic viscosity.}$$

5.3 Boundary Conditions

It is almost certain that the stratum of gas nearest to a solid body is in a very different state than the rest of the gas.

J. C. Maxwell, in *Phil. Trans.*, 170 (1879), 249–256, and *Scientific Papers* 2 (2003), 703–709.

Consider the viscous, incompressible NSE in a bounded domain Ω. Boundary conditions must be imposed on $\partial\Omega$ to have a completely specified problem. Let $\Gamma \subset \partial\Omega$ be a solid wall. The *no-slip condition* in fluid mechanics usually refers to the boundary condition $u = 0$ at a stationary wall, and we shall generally assume that this no-slip condition holds at solid walls. The no-slip condition is an excellent description of the behavior of a fluid near a wall in almost all situations. On the other hand, it is good to review where this condition originates (for more, see Day [24]), since there are situations where it should be modified. (We shall present a few examples later in this section.) There are also examples where what is really happening near a wall is still a mystery, such as is the case of *rapid water*.[41]

From first principles, then, what are the correct boundary conditions at a solid wall? The first boundary condition is easy:

$$\text{no penetration} \quad \Leftrightarrow \quad u \cdot \hat{n} = 0 \text{ on } \Gamma. \tag{5.3.1}$$

The tangential component is more complex. Navier proposed the following slip with friction condition:[42]

$$u \cdot \hat{\tau}_j + \beta \overrightarrow{t} \cdot \hat{\tau}_j = 0, \quad j = 1, 2. \tag{5.3.2}$$

NSEs can also be derived from the kinetic theory of gasses. In 1879 [70], Maxwell also derived the boundary condition at a solid wall for a continuum gas that comes from continuum limits of kinetic theory. The limit at a boundary gives exactly the boundary condition proposed by Navier (up to the accuracy of the linear approximation), where

$$\beta \sim \frac{\text{mean free path of molecules}}{\text{macroscopic length}}.$$

Thus, where the stresses are $O(1)$ the no slip condition

$$u \cdot \hat{\tau}_j = 0 \text{ on } \Gamma, \quad j = 1, 2,$$

[41]What is *rapid water* and why is it a mathematical and physical mystery? (Good question—try to find out!)

[42]This is often called the *Navier slip law* although it describes slip with resistance rather than free slip.

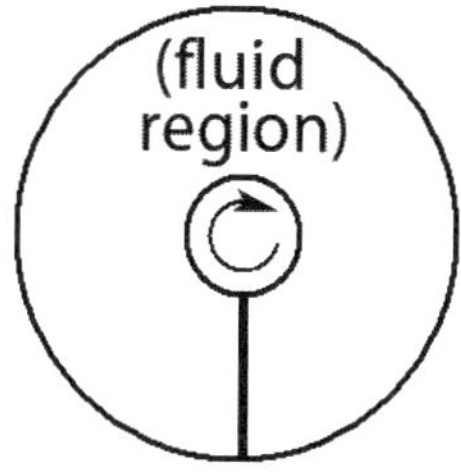

A line of ink is laid out in water between two circular pans. The inner pan is then rotated.

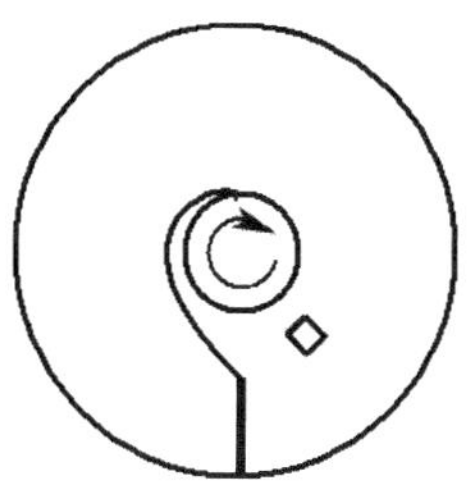

The ink line remains attached at the same point on the inner pan. Note the "boundary layer" structure in the flow.

If a rectangle of ink is laid out and the inner pan turned again, ...

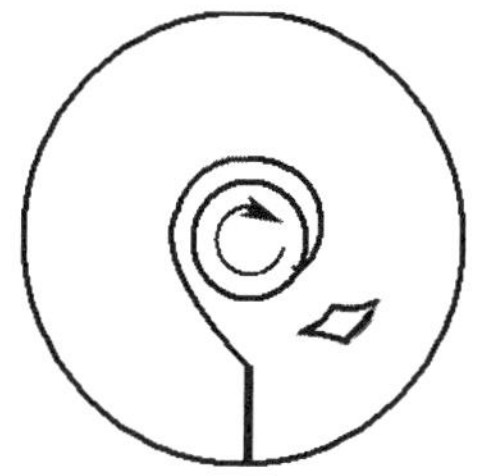

It deforms because of shear forces.

Figure 5.2. *Verifying no-slip at low stresses.*

is used and agrees well with experiments. If the boundary Γ is moving, this is modified to read

$$u = g \text{ on } \Gamma, \quad g = \text{ velocity of } \Gamma.$$

For example, a classic experiment (that is easy to repeat) verifies this no-slip condition. A cylinder is placed in a pan of water at rest. Next, a thin line of dye is inserted, connecting the boundary of the cylinder with the side of the pan. As the cylinder rotates, the line of dye twists to stay connected to both walls—as predicted by the no-slip condition. This is depicted in Figure 5.2.

There are cases in which the tangential stresses can become large or even infinite at points on Γ, such as in flow in domains with reentrant corners (flow over a step) and flows with discontinuous boundary conditions, such as in the driven cavity test problem.

Example 14 (driven cavity). The driven cavity problem is one of the most commonly used test problems in CFD. The flow domain is the unit square in two dimensions. There is no body force and the boundary conditions are $u = 0$ on three sides (bottom, left, and right).

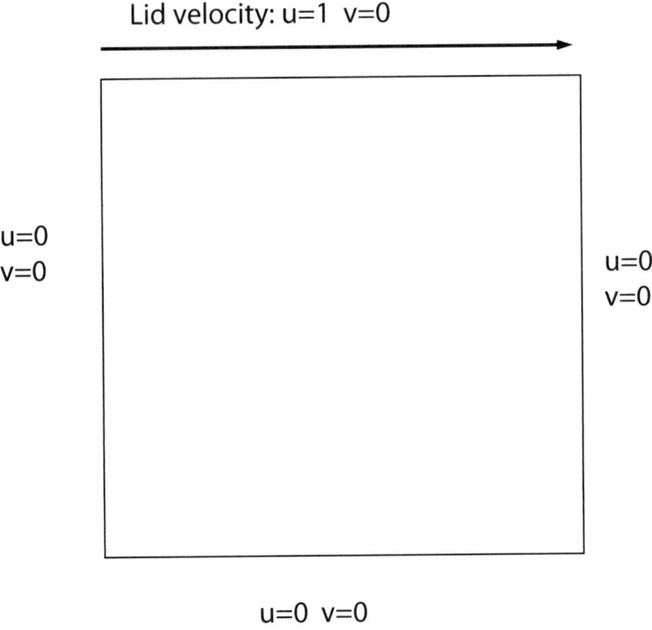

Figure 5.3. *Driven cavity domain and boundary conditions.*

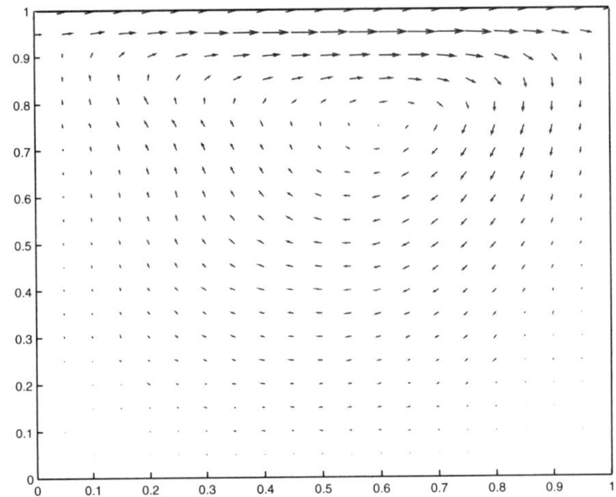

Figure 5.4. *An example of flow in the driven cavity.*

On the top side, the cavity lid, the boundary condition $u = (1, 0)$ is imposed (Figure 5.3). Because the geometry and boundary conditions are simple, it is easy to produce simulations of this flow and generally the flow circulates clockwise with some very complicated small eddies near the top corners (Figure 5.4). However, in computations generally the closer one zooms into the flow near those corners, the more inconclusive the simulations seem to be!

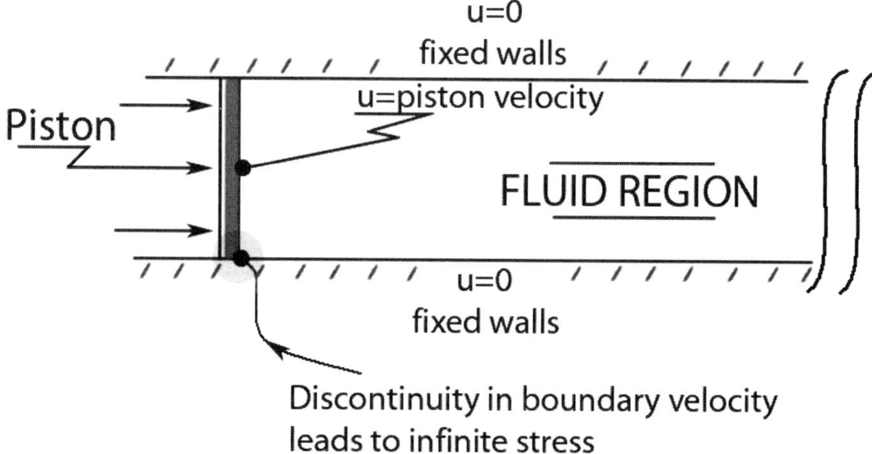

Figure 5.5. *Discontinuous boundary velocities induce infinite stress.*

Because the boundary conditions are discontinuous, any velocity must satisfy $u \notin H^1(\Omega)$. In other words, stress concentrates at the top corners to the point that the total force the flow exerts becomes infinite! Mathematically, it is not known if any solution exists in any reasonable sense (or even what sense would be reasonable). Physical experiments simulating the flow seem to invariably produce such high pressures at those corners that some water leaks out. These difficulties seem to all be due to imposing the simple no-slip boundary condition near the upper corners.[43] ∎

Flows in pistons, Figure 5.5, also have discontinuities in the boundary conditions and require modifications in the no-slip condition at those points. Another example that is of technological importance is coating flows. If a fluid flows over a solid surface, surface tension causes the leading edge of the flow to contact the surface at a nonzero angle creating, depending if the surface is wetting or nonwetting, a corner in the flow domain (Figure 5.6). One of the best test problems is the forward-backward step. This flow contains a laminar inflow condition, a small obstacle in the channel, and a long channel with outflow condition. The interesting features are the recirculation region of much slower flow, the reattachment point of the streamline separating this region from the main flow, this streamline itself, the eddies that are shed at high enough Reynolds number, and their period. A typical flow around the forward-backward step is given next. Note the recirculation region (whose length is a statistic of importance) and, at a high enough inflow velocity, eddies are shed from the step and flow downstream. This shedding is beginning in Figure 5.7. The period of shedding is also an interesting statistic. Flows around obstacles are also common in nature. The next photograph,[44] Figure 5.8, is one fun example. In this picture, three kayakers wait for their

[43] *Warning:* Since it is so easy to set up a numerical experiment with the driven cavity, it is often used. Unfortunately, it always seems to end up taking more time to arrive at a clear conclusion than when a test problem that is more clear mathematically and physically is used instead. Also, it seems that once one uses a test problem, it is destined to be repeated in many subsequent investigations, compounding the misery if badly chosen.

[44] Of A. Layton, taken by the author.

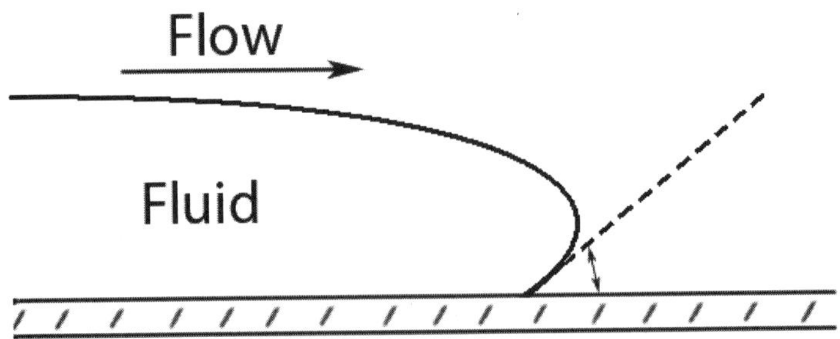

Figure 5.6. *A fluid flows across a surface.*

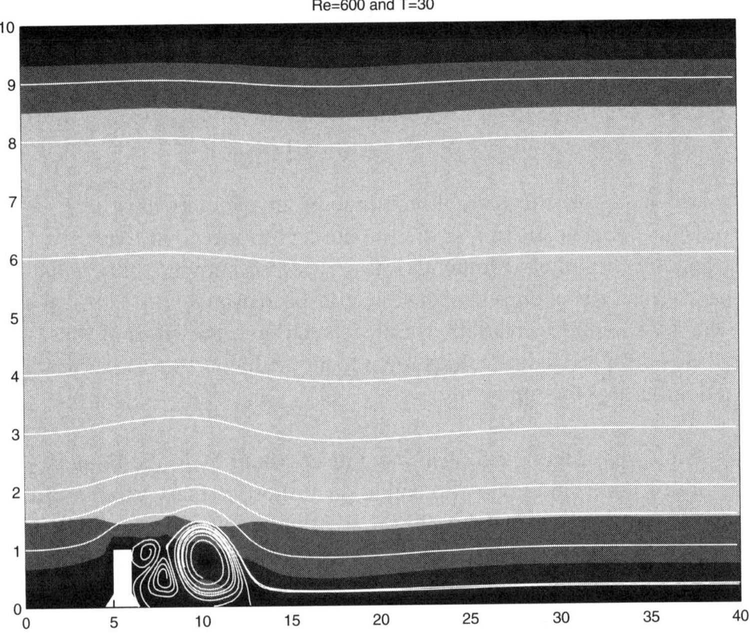

Figure 5.7. *A typical flow over a step.*

turn in the recirculating flow behind the obstacle. One kayaker in the foreground explores the Cauchy stress vector along the streamline separating the recirculation region from the faster flow in the free stream.

Imposing the no-slip condition at this leading edge leads to nonphysical predictions because of nonphysical stress accumulation at those corners. In such problems, the no-slip condition must be replaced by slip with friction near the points of large stresses. The slip with friction condition is often called *Navier's slip law*. There are other problems with corners where it is not clear if modifications are physically appropriate. The forward-backward step problem (depicted in the Figure 5.9) is a great test problem (being much harder than it seems) and one such example.

Figure 5.8. *Exploring the Cauchy stress vector.*

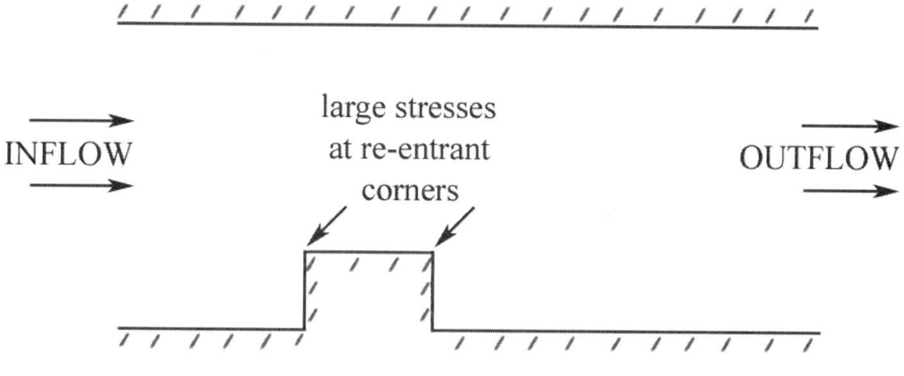

INFLOW

large stresses
at re-entrant
corners

OUTFLOW

Figure 5.9. *Geometry of the forward-backward step.*

5.4 The Reynolds Number

The use of the Reynolds number as an independent variable is an application of a basic truth, not just a useful convention for a handy diagram.
N. Rott, from [82].

One important question in fluid dynamics is how to use information about flows in experiments to understand flows of different fluids occurring on different scales of motion and at different velocities. The answer to this question uses techniques called dynamic similarity, dimensional analysis, and nondimensionalization and always involves a special parameter called the Reynolds number.

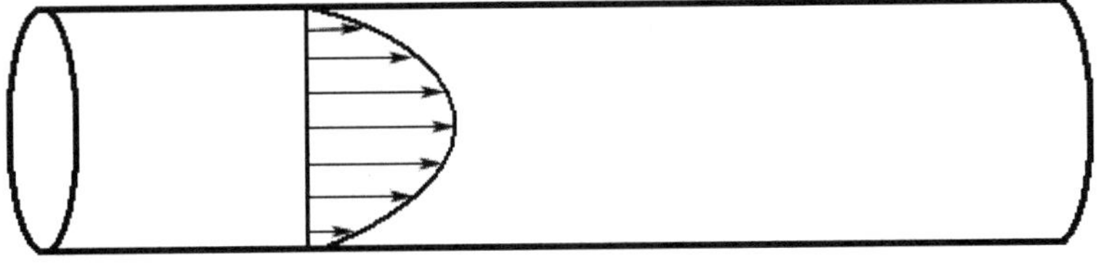

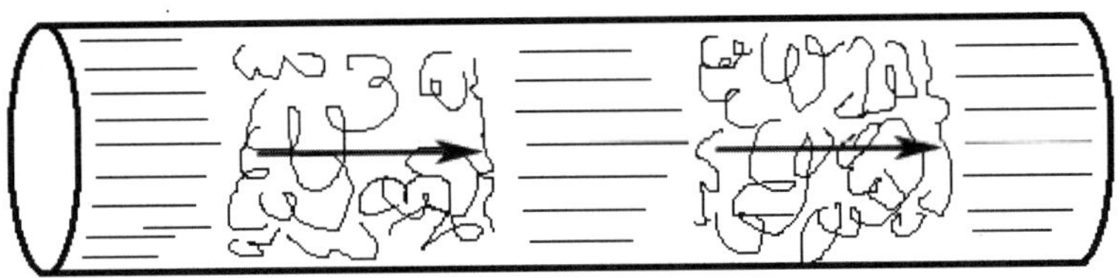

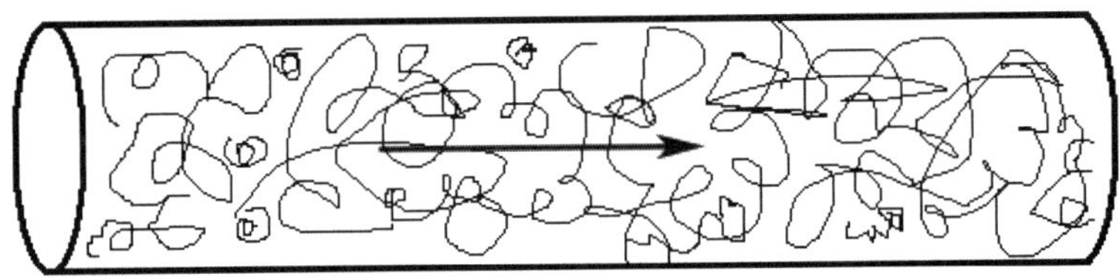

Figure 5.10. *Flow in pipes at increasing Reynolds numbers.*

Example 15 (laminar and turbulent flow in pipes). One of the central problems in early fluid mechanics was to understand flows in pipes. Experiments in pipes were easy to perform: a glass tube is filled with a fluid and attached to a pump. The flow is visualized easily by shining a light through the pipe. Then the flow is observed varying (i) the pipe's diameter, (ii) the flow rate (by controlling the pump), (iii) the liquid in the pipe (e.g., substituting glycerin for water), and (iv) the fluid's density (e.g., by varying the temperature). Three typical flow patterns (sketched in Figure 5.10) are seen:

- The flow can be laminar in the pipe,
- The flow has turbulent regions (called turbulent slugs) separated by laminar regions, and
- The flow is fully turbulent. ∎

The scientific problem was to take these observations and make a prediction about an industrial flow occurring for vastly different flow rates, pipe diameters, liquids, and temperatures. Osborne Reynolds used dimensional analysis to solve exactly this problem. This section shows one application of dimensional analysis to the NSE: the derivation of the famous Reynolds number.

As another example of current importance, if aerospace engineers and designers study the flow of air around an airplane, typically the airplane is scaled down by a factor of, say, 30–50 and the model is put in a wind tunnel. It is a real problem though to adjust the results of the wind tunnel to the full-scale airplane. This method of adjustment is called dynamic similarity and it is due to Reynolds building on the work of Froude in 1883.

Consider the incompressible NSE

$$\nabla \cdot u = 0 \text{ and } \partial u/\partial t + u \cdot \nabla u + \nabla(p/\rho_0) = (\mu/\rho_0)\Delta u.$$

Recall that μ is the dynamic viscosity and $\nu := \mu/\rho_0$ is the kinematic viscosity. The flow occurs in a domain Ω and the flow geometry has some characteristic length L. For example, L can be the length of the airplane. Thus, we can geometrically scale the domain to get a rescaled domain $\Omega^* = \Omega/L$. Ω and Ω^* are geometrically similar. Now rescale the remaining variables as follows:

$$\begin{aligned}
&x^* := x/L \quad L := \text{ a reference length,}\\
&u^* = u/V, \quad V \text{ a reference speed,}\\
&t^* = V\,t/L,\\
&p^* := [p - p_0]/\rho^* V^2, \quad p_0 \text{ a reference pressure, } \rho_0 \text{ a reference density.}
\end{aligned}$$

V is a reference speed, for example, the free stream speed away from the model airplane. If we denote partial derivatives with respect to the rescaled, dimensionless variables by $\frac{\partial}{\partial x^*}$ and so on, then the chain rule implies[45]

$$\frac{\partial u^*}{\partial x^*} = \frac{L}{V}\frac{\partial u}{\partial x}, \quad \text{for example.}$$

Then the NSE for incompressible flows in the rescaled variables become

$$\frac{\partial u^*}{\partial t^*} + u^* \cdot \nabla u^* = -\nabla p^* + \left(\frac{\mu}{\rho_0\, V L}\right) \Delta^* u^*.$$

Thus the flow around the original airplane will be dynamically similar to the flow around the airplane of length one if the original and rescaled NSE are the same.

Definition 21 (dynamic similarity). *Flows in similar geometry Ω and $\Omega^* := \Omega/L$ are dynamically similar if the parameters of the flow are such that the quantities of the flows*

$$\text{Re} := \frac{\rho\, V L}{\mu}$$

coincide. The dimensionless parameter Re *is called the Reynolds number.*

[45] $\frac{\partial u^*}{\partial x^*} = \frac{\partial}{\partial x^*}(u/V) = \frac{1}{V}\frac{\partial u}{\partial x}\frac{\partial x}{\partial x^*} = \frac{1}{V}\frac{\partial u}{\partial x}L.$

The Reynolds number Re describes different flow properties. It represents the ratio of the inertial forces to viscous forces:

$$\text{Re} = \text{inertia terms/viscous terms.}$$

If Re is close to 0, then viscous forces dominate inertial forces, as occurs in a highly viscous fluid moving slowly. For very large Re the viscous forces can be neglected. This often occurs in flows of gases. The Reynolds number occurs in all flow settings.

Taking advantage of dynamic similarity in practical settings can be challenging. For example, if an airplane model is reduced by 30 times smaller than the original airplane and is put in a wind tunnel, the reduction will reduce the Reynolds number by a factor of 30 from the original airplane. Thus, the other conditions in the wind tunnel must be adjusted to compensate—for example, increasing the speed of the wind tunnel by a factor of 30. This is not practical, so the only other possibility is to change the gas, thereby changing the reference density and viscosity. Conventional wind tunnels try this by increasing the pressure. This generally is not sufficient, so wind tunnel tests have traditionally been performed at Reynolds numbers only 20% of the Reynolds number of the true flow geometry. To put this difficulty in a concrete setting, there was a recent European project to try to achieve the correct Reynolds number in wind tunnel experiments. On top of increasing the gas pressure, nitrogen had to be substituted for air, and the nitrogen had to be cooled down to -183 degrees C. The construction of the wind tunnel is estimated at $390 million and each run is estimated to require 50 megawatts of electric power (this is roughly the electricity usage of a small town). Thus, you can see that wind tunnel tests are very expensive and the conclusions are often uncertain. In fact, the uncertainty of wind tunnel test has been one driving force behind the development of computational fluid dynamics.

It is illuminating to consider a few simple flow problems and their associated Reynold's number. From (5.4.1) it seems that for many flow problems the case of high Re is the one of interest!

$$\begin{array}{lll} \text{Model airplane} & 1\text{m/s} & \text{Re} \sim 7 \times 10^4 \\ \text{Subcompact car} & 3\text{m/s} & \text{Re} \sim 6 \times 10^5 \\ \text{Small airplane} & 30\text{m/s} & \text{Re} \sim 2 \times 10^7 \end{array} \tag{5.4.1}$$

There are two interesting limiting cases in the incompressible NSEs: when Re approaches zero and infinity. The case when Re $\to 0$ is understood; it is called *creeping flow* or *Stokes flow*. When Re $\to 0$ the characteristic velocity of the flow can be considered to approach 0. In physical mathematics, a small number usually means that all quadratic quantities in that number are negligible. Hence in this limit the $v \cdot \nabla v$ term in the NSE drops out of the equations and we have the following system, called the Stokes problem:

$$\begin{aligned} -\nu \Delta u + \nabla p &= 0 \quad \text{in the flow domain,} \\ \nabla \cdot u &= 0 \quad \text{in the flow domain.} \end{aligned}$$

The Stokes problem requires boundary conditions, most often $u = 0$ on the boundary of the domain, and some normalization condition on the pressure such as requiring the pressure to have mean value zero.

The case when Re becomes very large is a very common case (see the last table) and the least understood because it is linked to the question of turbulence! Even before the flow becomes turbulent, interesting flow structures, such as the von Karman vortex street (which begins to form around Re = 100) in Figure 6.5, are formed.

5.5 Boundary Layers

I have posed myself the task to do a systematic research about the laws of motion for a fluid in such cases when friction is assumed to be very small.
L. Prandtl, International Congress of Mathematicians, 1904, quoted in [29].

When the complete mathematical problem looks hopeless, it is recommended to enquire what happens when one essential parameter of the problem reaches the limit zero.
L. Prandtl, in *Mein Weg zu Hydrodynamischen Theorien*, Physikalische Blätter, Deutsche Physikalische Gesellschaft, Berlin, 1948, p. 1606.

They (paradoxes) can be traced to the use of plausible arguments,.... Among these are the arguments that "small causes produce small effects" and that "symmetric causes produce symmetric effects."
G. Birkhoff, in *Hydrodynamics: A Study in Logic, Fact and Similitude*, Princeton, 1950.

When the Reynolds number is large, $\frac{1}{\text{Re}}$ is small. Thus, if the velocities' second derivatives are $O(1)$, $|\text{Re}^{-1}\Delta u|$ must be small over most of the region. Yet, dropping this term leads to the Euler equations for inviscid flows which are insufficient to predict many flow quantities:

$$u_t + u \cdot \nabla u + \nabla p = f \text{ and } \nabla \cdot u = 0 \text{ in } \Omega.$$

Although the Euler equations have many interesting features, they are unable to predict, for example, the drag experienced by an object moving through the air (which is very nearly inviscid).[46] (This was noted by d'Alembert and is known as the d'Alembert paradox.)

I must confess that I do not know how the resistance of fluids can be explained theoretically, because the theory gives zero resistance...."
d'Alembert, 1768, "Paradoxe proposé aux géomètres sur résistance des fluides," in *Opuscules mathématiques*, Vol. 5, 34th Memoir, Paris.

This was one of the great mysteries of fluid mechanics that wasn't resolved until the 20th century! The solution to this puzzle was found by Ludwig Prandtl in 1904.[47] In his work, he showed that at higher Reynolds numbers, viscous effects were confined to the boundary layer, a thin region along a body immersed in the fluid, and in the wake behind the body. Although in most of the flow region, the velocity satisfies the Euler equations (very nearly), it is precisely the thin region along the body and the wake which determines

[46] The case of *lift* is more complicated. Since lift can be primarily caused by pressure and the Euler equations include pressure, there are settings in which lift can be accurately predicted by solutions of the Euler equations.

[47] Prandtl presented his idea to the third International Congress of Mathematicians in Heidelberg in 1904. Arguably, it was the greatest advance ever in the mathematical understanding of fluid motion and was largely ignored at the congress!

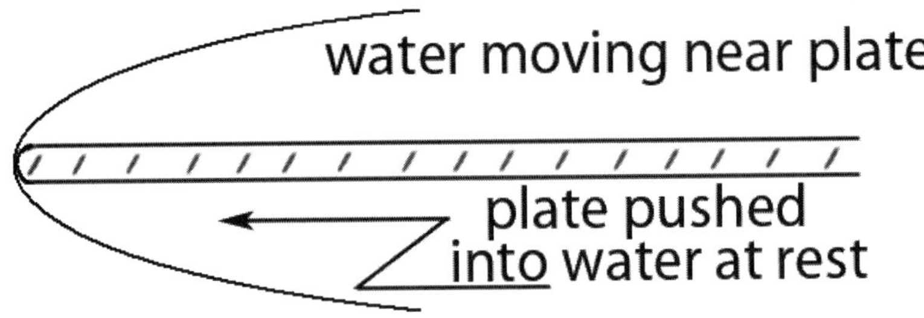

Figure 5.11. *A boundary layer flow.*

quantities like drag. Thus, knowledge of the width of these regions and variation of the flow variables inside them is absolutely necessary for generating finite element meshes which can be used to predict correctly such quantities.

Prandtl's estimate of the width of a laminar boundary layer is absolutely fundamental to generating an appropriate mesh for flow problems: the mesh typically must be considerably finer near walls than away from walls. In particular, even if only the global flow field is sought, three mesh lines must be placed close enough to walls to be in the boundary layer. If stresses on the wall are needed, even more meshlines are needed inside the layer. Often, the elements are long and thin (being much smaller normal to a wall than along a wall) and a coarser mesh can be used for the pressure.

To give Prandtl's idea some mathematical structure, consider a two-dimensional flow $(u(x, y), v(x, y), 0)$ along an infinite flat plate. This is a simplification and idealization of a real flow: a finite plate is pushed into a fluid or a fluid flows past a stationary flat plate with uniform velocity (Figure 5.11). The flow away from the leading edge is observed. The next figure gives a sketch of the experimental results. Away from very close to the plate, the flow looks uniform and laminar. The flow very close to the plate can be visualized by putting a very thin wire in the flow that gives off bubbles. These bubbles flow with the fluid and so their displacement from the wire is a visualization of the velocity vector!

This is described by the schematic which exaggerates the width of the region near the plate region. Suppose that away from the plate the flow is laminar and (up to the accuracy of our picture) is given by

$$u = V, \ v = 0 \ \text{ satisfying the Euler equations.}$$

Reverse the above picture and imagine the plate to be long and at rest with the fluid flowing over it with nearly uniform velocity. The reasoning of Prandtl was that if $(u(x, y), v(x, y))$ must go from $(0, 0)$ along the wall to a free stream value $u(x, y) = V, v(x, y) = 0$ a certain distance δ away from the wall, then (u, v) must be very nearly as depicted in Figure 5.12.

The "boundary layer" is the $O(\delta)$ transition region near the wall. Inside the boundary layer, (u, v) must satisfy the incompressible NSE, which we assume here (to simplify the

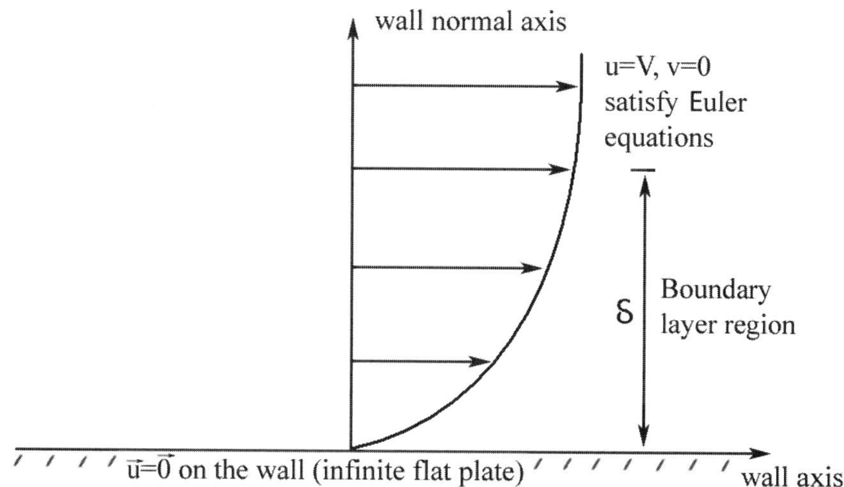

Figure 5.12. *Depiction of a boundary layer.*

presentation) to be steady as well: $\vec{u} = (u, v)$ satisfies

$$\begin{aligned} u\,\frac{\partial u}{\partial x} + v\,\frac{\partial u}{\partial y} + \frac{\partial p}{\partial x} - \mathrm{Re}^{-1}\Delta u = 0 \\ & \qquad \text{for } 0 < y < \delta \\ u\,\frac{\partial v}{\partial x} + v\,\frac{\partial v}{\partial y} + \frac{\partial p}{\partial y} - \mathrm{Re}^{-1}\Delta u = 0, \end{aligned}$$

and $u(x, 0) = v(x, 0) = 0$, $u(x, \delta) = V$, $v(x, \delta) = 0$. Consulting the last figure, it is easy to estimate the order of magnitude of the various terms in the NSE. Indeed, across the layer (i.e., in the y-direction) u changes from $V = O(1)$ to 0 in δ distance. Thus, $\frac{\partial u}{\partial y} \doteq O(\delta^{-1})$, in the layer. This estimate can be puzzling at first. It is simply

$$\begin{aligned} \frac{\partial u}{\partial y} &\doteq \frac{\text{change in u}}{\text{distance over which change occurs}} \\ &= \frac{V - 0}{\delta} = O(\delta^{-1}). \end{aligned}$$

Similarly, $\frac{\partial^2 u}{\partial y^2} = O(\delta^{-2})$. At the edge of the layer $u = O(1)$ (i.e., $u = V$ when $y = \delta$), and u changes little in the x-direction. Thus, we can estimate

$$u, \frac{\partial u}{\partial x}, \frac{\partial^2 u}{\partial x^2} = O(1) \text{ in the layer.}$$

By the continuity equation $\frac{\partial u}{\partial x} + \frac{\partial v}{\partial y} = 0$. Thus, $\frac{\partial v}{\partial y} = O(1)$, in the layer. Since $v(x, 0) = 0$ and $\frac{\partial v}{\partial y} = O(1)$ in the layer, the mean value theorem implies

$$v = O(\delta) \text{ in the layer.}$$

Similarly,

$$\frac{\partial v}{\partial x} \text{ and } \frac{\partial^2 v}{\partial x^2} = O(\delta) \text{ in the layer.}$$

Since $\frac{\partial v}{\partial y} = O(1)$ in the layer, it also follows that

$$\frac{\partial^2 v}{\partial y^2} = O(\delta^{-1})$$

in the layer. The pressure is not constrained on the wall. Thus, in the layer we expect $p = O(1)$ and $\frac{\partial p}{\partial x} = O(1)$.

With these sizes of terms, reconsider the momentum equation. In the layers

$$\underbrace{u\frac{\partial u}{\partial x}}_{0(1)\cdot 0(1)} + \underbrace{v\frac{\partial u}{\partial y}}_{0(\delta)\cdot 0(\delta^{-1})+} + \underbrace{\frac{\partial p}{\partial x}}_{+O(1)} \underset{=}{=} \mathrm{Re}^{-1}\underbrace{\left(\frac{\partial^2 u}{\partial x^2} + \frac{\partial^2 u}{\partial y^2}\right)}_{\mathrm{Re}^{-1}(O(1)+O(\delta^{-2}))}.$$

For $\mathrm{Re}^{-1} << O(1)$ the $\mathrm{Re}^{-1}\,\frac{\partial^2 u}{\partial x^2}$ term is negligible. Thus, the x-momentum equation reduces to:

$$O(1) = O(\mathrm{Re}^{-1}\delta^{-2}).$$

Thus, the thickness δ of the boundary layer is

$$\delta = O(\mathrm{Re}^{-1/2}).$$

This can be very small and implies that a correspondingly small mesh *must* be used near walls for the velocity variables.

Consider the y-momentum equation. In the layer,

$$\underbrace{u\frac{\partial v}{\partial x}}_{0(1)\cdot 0(\delta)} + \underbrace{v\frac{\partial v}{\partial y}}_{0(\delta)\cdot 0(1)} + \underbrace{\frac{\partial p}{\partial y}}_{+\frac{\partial p}{\partial y}} \underset{=}{=} \mathrm{Re}^{-1}\underbrace{\left(\frac{\partial^2 v}{\partial x^2} + \frac{\partial^2 v}{\partial y^2}\right)}_{\mathrm{Re}^{-1}(O(\delta)+O(\delta^{-1}))}.$$

The term $\mathrm{Re}^{-1}\,\frac{\partial^2 v}{\partial x^2}$ is negligible and $\mathrm{Re}^{-1}\,\frac{\partial^2 v}{\partial y^2} = 0(Re^{-1}\delta^{-1}) = O(\delta)$. This gives

$$\frac{\partial p}{\partial y} = O(\delta) \text{ in the layer.}$$

Since $\frac{\partial p}{\partial y} = O(\delta)$ in the layer, the change in pressure across the layer is $O(\delta^2)$-negligible. Interestingly, this implies that a fine mesh is not needed for the pressure variables in laminar boundary layers. The pressure is very close to the free stream pressure. Thus, typical meshes used for boundary layer flows tend to be very coarse for the pressure, coarse for the velocity if the streamwise direction, and very fine in the wall normal direction.

5.6 An Example of Fluid Motion: The Taylor Experiment

Wisdom is the daughter of experience.
L. da Vinci, in: The Notebooks of Leonardo da Vinci, Forst III 14v; Macurdy 80.

History of science reveals to us two kinds of phenomena, opposite as it were: at times, simplicity is hidden behind apparent complexities; at other times, on the contrary, we find behind apparent simplicity hides extremely complicated realities.
H. Poincaré, in *Hypothesis in Physics*, quoted in [21].

One of the most important examples of fluid motion is flow between rotating cylinders, studied first by Couette and extensively by Taylor. This is now often called the Taylor experiment because the definitive experiments were conceived and carried out by G. I. Taylor. His initial experimental results were, in fact, greeted with skepticism partially because they were so dramatic and partially because the first attempts to replicate his results failed because the experimental technique used was not up to that of Taylor. In Taylor's experiments, two very long cylinders are set up to have the same center, and the space between is filled with a liquid. Each can rotate. The difference between their angular velocities $\omega = |\omega_{outer} - \omega_{inner}|$ is the key control parameter in the experiment (Figure 5.13).

There are various ways to visualize the flow. For example, with clear cylinders, reflective particles can be suspended in the liquid and a light shone through. The fluid trajectories will then be traced out by the particles' reflections and can be photographed. (Many interesting photographs of this experiment are in van Dÿk's book [28].)

Here is a synopsis of Taylor's observations:

- For ω small, the trajectories were simple circles—each cylindrical shell of fluid was undergoing a rigid body rotation.
- As ω increased, the particles' speeds of rotation increased until ω reached a critical value ω^*.
- For $\omega = \omega^*$, the critical value, the flow reorganized itself very quickly. The new flow was composed of three cells—each a superposition of a rotation and a rolling motion inside the cell (Figure 5.14). (These cells are now called Taylor cells.)
- As ω increases beyond ω^* the speed of rotation inside each cell increases but the same cell structure remains.
- This persists until the next critical value of ω is reached. At this value the flow reorganizes into five Taylor cells.
- At some point, the observed steady Taylor cells lose their stability (as a solution of the time-dependent NSE) and the observed flow becomes time periodic: the borders between the cells become wavy.
- At still higher ω the flow becomes very complex: first a superposition of periodic motions with different periods, then laminar cells alternating with turbulent cells, then fully turbulent.

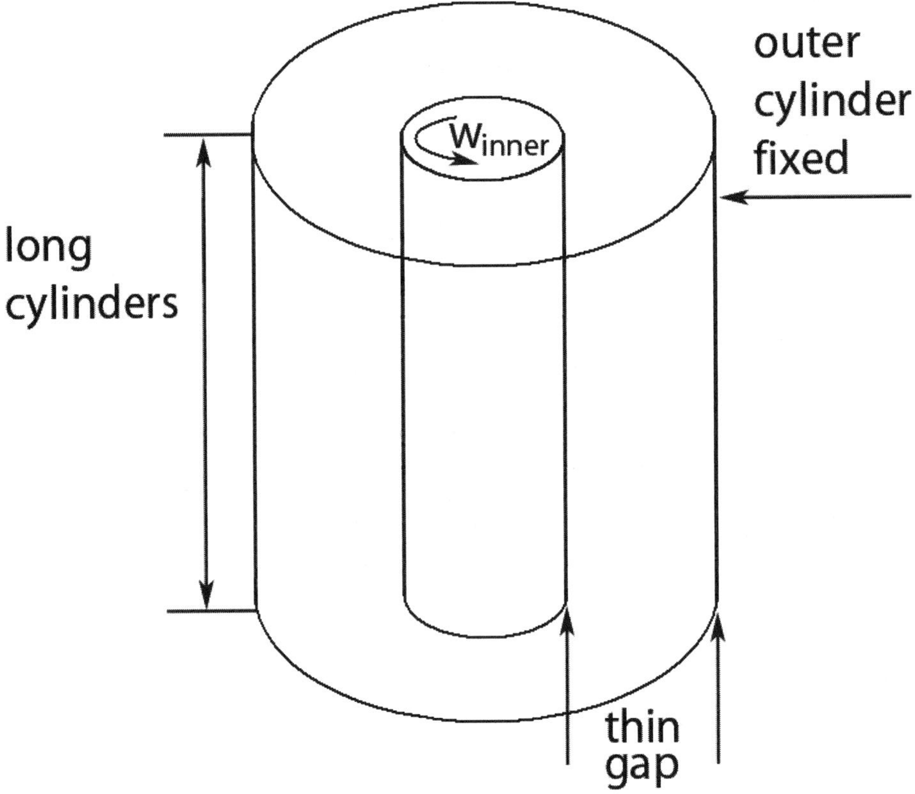

Figure 5.13. *Setup of counter rotating cylinders.*

The pictures in the albums of fluid motion [28], [99] are indispensable for understanding this description! For low values of $\lambda = \omega$ this description is often summarized in a diagram (Figure 5.15).

5.7 Remarks on Chapter 5

The best presentation of mathematical fluid dynamics is the famous 1959 article by Serrin [87]. Other interesting presentations of mathematical fluid dynamics are in the books by Hughes and Marsden [54] (which is, unfortunately, out of print), Batchelor [12], Dautry and Lions [23], and Meyer [71]. The fundamental reference in boundary layer theory is the current edition of Schlichting's book, e.g., [86]. Our schematics of various flows are poor cartoons compared to observations of the real thing, as in the books by Turek, Kilian and Turek [99] and van Dyke [28]. For an interesting scientific history of fluid dynamics, see the books by Tokaty [97] and Darrigol [22] and Batchelor's biography of G. I. Taylor [13]. Pressure was introduced as a mathematical variable in fluid models in the first book on mathematical fluid mechanics, *Hydrostatics*, by Archimedes [3].

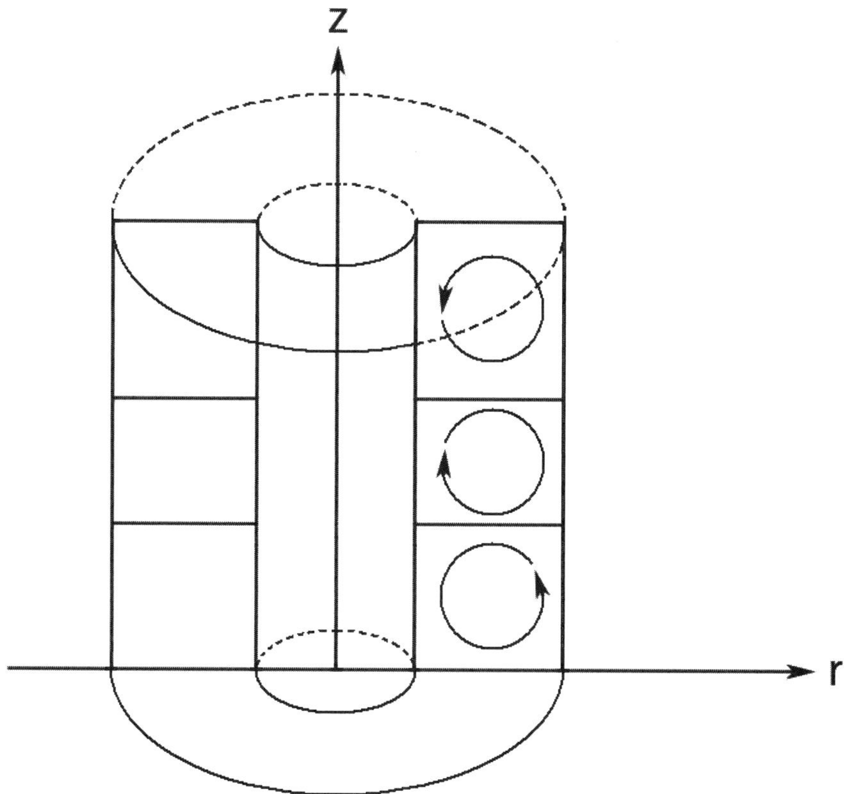

Figure 5.14. *Three Taylor cells.*

The operator $\phi \to \frac{\partial \phi}{\partial t} + u \cdot \nabla \phi$ is known as the material derivative and denoted $\frac{d\phi}{dt}$ or $\frac{D\phi}{Dt}$:

$$\frac{d\phi}{dt} := \frac{\partial \phi}{\partial t} + u \cdot \nabla \phi.$$

Its interpretation is interesting. If, for example, ϕ denotes the temperature, $\frac{\partial \phi}{\partial t}$ denotes the rate of change of the mercury level in a thermometer sitting at a fixed point in space. On the other hand, if the thermometer is stuck on a fluid particle that moves with the fluid, the rate of change of temperature experienced by the moving thermometer is the material derivative $\frac{d\phi}{dt}$. The material derivative naturally occurs in continuum models of processes associated with particles or groups of particles, such as momentum and acceleration.

The foundation of fluid (and even continuum) mechanics is the stress principle of Cauchy. The stress principle of Cauchy is simplicity itself: stress forces are contact forces. We take a statement of it following Serrin [87].

Definition 22 (Cauchy stress principle). *Upon any imagined closed surface S there exists a distribution of stress vectors $\overrightarrow{t}$, that is, for every point on the surface and every time there*

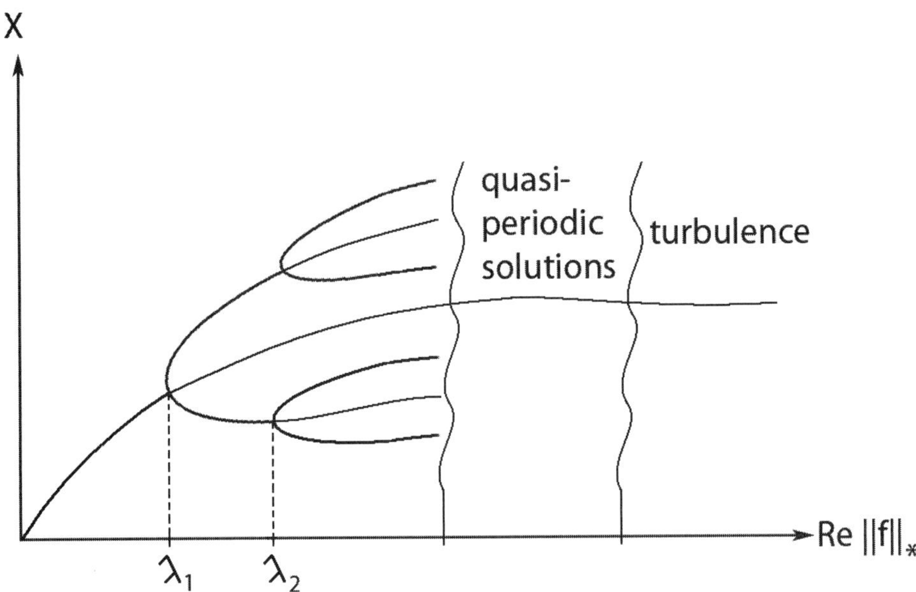

Figure 5.15. *Schematic of flow between rotating cylinders.*

corresponds a vector. The resultant and momentum of these stress factors are equivalent to those of the actual forces of material continuity exerted by the materials outside S upon the material inside S. The stress vector $\vec{t}$ *is assumed to depend only on the position* (x, t) *and the orientation of the tangent plane to the surface,* $\widehat{n}$*. Thus, if* $\widehat{n}$ *denotes the outward unit normal to S, then* $\vec{t} = \vec{t}\,(x, t; \widehat{n})$.

Before Cauchy, a whole century of brilliant geometers had struggled with the idea of stress and strain in continuous media and very complicated and usually incorrect ways. There are other ways to state the stress principle of Cauchy. Let us quote one more, given by Truesdell [98].

Definition 23 (Cauchy stress principle). *Upon any smooth, closed, orientable surface S, be it an imaginative surface within the body or the bounding surface of the body itself, there exists an integrable field of traction* $\vec{t}$*, with the same resultant and moment to the action exerted by the matter exterior to S and contiguous to it on the interior of S.*

Our presentation of the stress tensor is not far from that in the original and beautiful paper by George Stokes, published when he was only 26 years old. In his paper, the essence of this stress tensor was captured in his description, which is paraphrased as follows:

> The difference between the pressure on a plane in a given direction passing through any point P of a fluid in motion and the pressure which would exist in all directions about P if the fluid in its neighborhood were in a state of relative equilibrium depends only on the relative motion of the fluid immediately about

P, and the relative motion due to a rotation may be limited without affecting the above pressure differences.

The precise and rigorous mathematical realizations of this description are the following four Stokes postulates:

Condition 2, continuity. Π is a continuous function of the deformations tensor $\mathbf{D}$ and is independent of all other kinetic quantities. It can possibly depend upon the fluids thermodynamic state however.

Condition 3, spatial homogeneity. Π does not depend explicitly upon the position x.

Condition 4, isotropy. There is no preferred direction in space.

Condition 5, no tangential stress at equilibrium. When $\mathbf{D} = 0$, Π reduces to $-p\mathbf{I}$.

The consequences of the four Stokes postulates form the mathematical basis of our understanding of the internal forces in a fluid.

5.8 Exercises

Exercise 51. (a) *Show that the force exerted on a fluid volume by the rest of the fluid is exactly the negative of the force exerted on the rest of the fluid by the fluid volume. This is another important result of Cauchy. (**Hint:** This is very easy once Cauchy's theorem is known and considerably harder from first principles.)*

(b) *If the volume is the entire fluid enclosed by a fixed wall, find an expression for the force of the fluid on the wall.*

Exercise 52 (flow rate $\times$ cross sectional area = constant). *Use $\nabla \cdot u = 0$ and integration over the flow domain to show that da Vinci's description of flow through a syringe (that flow rate $\times$ cross sectional area is constant) was correct.*

Exercise 53 (the fluid acceleration). *Fill in the details of the derivation of the formula for the acceleration. There are three ways this is typically approached: the transport theorem, the chain rule, and a direct calculation using Taylor expansions to two terms (i.e., linear approximations). We shall use the last.*

Consider the path of a fluid particle. Suppose that at time t it is at the point x in space. Its velocity is then $u(x,t)$. At later time $t + \Delta t$ it has moved along the particle path. If the path is smooth and Δt is small, the point it has moved to is approximately $x + u(x,t)\Delta t$ $(+O(\Delta t^2))$.

(a) *Draw a schematic of a particle path. Indicate on it the two points and the velocity vector. The velocity at this second point is $u(x + u(x,t)\Delta t, t + \Delta t)$. The acceleration is the change in velocity per unit change in time. Thus,*

$$a(x,t) = \lim_{\Delta t \to 0} \frac{u(x + u(x,t)\Delta t, t + \Delta t) - u(x,t)}{\Delta t}.$$

(b) *Recall the Taylor expansion:* $u(x+h) = u(x) + h \cdot \nabla u + O(|h|^2)$. *Use this expansion in the above limit to show that the fluid's acceleration is*

$$a = \frac{\partial u}{\partial t} + (u \cdot \nabla)u.$$

Another approach is via the chain rule.

(c) *Reconsider the above difference quotient. Add and subtract terms and write the difference quotient as*

$$\frac{u(x + u(x,t)\Delta t, t + \Delta t) - u(x,t)}{\Delta t} = A + B,$$

where as $\Delta t \to 0$, $A \to \frac{\partial u}{\partial t}$, $B \to (u \cdot \nabla)u$.

This proof is equivalent to using the chain rule!

Exercise 54 (another formula for acceleration). *Show*

$$a = \frac{\partial u}{\partial t} + (\nabla \times u) \times u + \frac{1}{2}\nabla(u \cdot u).$$

Exercise 55 (the vorticity transport equation). *Take the curl of Euler equations (see section 5.2) and derive the following equation for an incompressible, inviscid fluid's vorticity:*

$$\omega := \nabla \times u \textit{ satisfies } \frac{\partial \omega}{\partial t} + u \cdot \nabla\omega = \omega \cdot \nabla u + \rho_0^{-1} \nabla \times f.$$

Explain why $\omega \cdot \nabla u = 0$ *for a two-dimensional flow. How does the vorticity transport equation simplify in two dimensions?*

Exercise 56 (yet another vector identity). *Show that*

$$u \cdot \nabla u \cdot u = \frac{1}{2}\nabla \cdot (u|u|^2) - \frac{1}{2}|u|^2 \nabla \cdot u.$$

Exercise 57 (more on the vorticity equation). *Take the curl of Euler equations and derive the following equation for an incompressible, inviscid fluid's vorticity:*

$$\omega := \nabla \times u \textit{ satisfies } \frac{\partial \omega}{\partial t} + u \cdot \nabla\omega = \omega \cdot \nabla u + \rho_0^{-1} \nabla \times f.$$

Explain why $\omega \cdot \nabla u = 0$ *for a two-dimensional flow. How does the vorticity transport equation simplify in two dimensions?*

Exercise 58 (alternate forms of the viscous term). *Show that if* $\nabla \cdot u = 0$,

$$2\mu\nabla \cdot \mathbf{D}(u) = \mu \, \Delta u.$$

Exercise 59 (skew-symmetry). *Use* $u \cdot \nabla u \cdot u = \frac{1}{2}\nabla \cdot (u|u|^2) - \frac{1}{2}|u|^2 \nabla \cdot u$ *(from Chapter 3) and show* $\int_\Omega (u \cdot \nabla u) \cdot u \, dx = 0$ *if*

$$\nabla \cdot u = 0 \quad in \ \Omega \ and \quad u \cdot \hat{n} = 0, \quad on \ \partial\Omega.$$

Exercise 60 (Reynolds averaging). *Define the Reynolds averaging operator:*

$$\overline{u}(x) = \lim_{T \to \infty} \frac{1}{T} \int_0^T u(x,t)\, dt.$$

Average the incompressible NSE in time and show that provided the fluid velocity is bounded $\overline{u}$ *satisfies*

$$-\nu \Delta \overline{u} + \overline{u} \cdot \nabla \overline{u} + \nabla \overline{p} + \nabla \cdot \mathbf{R} = \overline{f},$$

where $\mathbf{R}(u) = \overline{u\,u} - \overline{u}\,\overline{u}$.[48] ***Hint:*** $u \cdot \nabla u = \nabla \cdot (uu)$.

Exercise 61. *Show that for a perfect fluid when* $p > 0$ *the surface stresses compress the fluid inside and when* $p < 0$ *the surface stress expands the fluid inside.*

Exercise 62 (Stress in shear flow). *Reconsider the case of shear flow (see Chapter 2). If the top* $(y = 1)$ *and bottom* $(y = 0)$ *plates are a unit distance apart and the top moves with unit velocity, the pressure is constant and velocity is given by*

$$u(x, y, z) = (y, 0, 0).$$

It is reasonable that the force this flow exerts pushing up on the top plate will be different from the force it exerts in the direction the plate is moving. This exercise is to verify this intuition by calculating these force directly from the explicit solution for the velocity and pressure.

(a) *Find the deformation tensor and stress tensor. Find the stress tensor at the top wall. In particular, find the force the flow exerts at the top wall in the normal direction to the wall (pushing up), and verify that in this case it is the pressure. Find the force exerted at the top wall in the direction the wall is moving. Sketch the geometry, velocity, and Cauchy stress vectors.*

(b) *Verify that forces depend on orientation by checking that the Cauchy stress vector at mid-channel is different on the* $x - y$, $y - z$ *and* $x - z$ *planes.*

Exercise 63 (a compatability condition on boundary velocities). *Suppose* u *satisfies the incompressible NSEs with the boundary condition*

$$u = g \quad on \ \Gamma.$$

Show that g *must satisfy the compatibility condition*

$$\int_\Gamma g \cdot \hat{n}\, ds = 0.$$

***(Hint:** Use* $\nabla \cdot u = 0$ *and the divergence theorem.)*

[48]These are known as the Reynolds averaged NSE and $\mathbf{R}(u) = \overline{u\,u} - \overline{u}\,\overline{u}$ is called the Reynolds stress tensor.

Exercise 64. *Consider the Stokes problem under the boundary conditions* (3.1), (3.2). *Find its variational formulation.*

Exercise 65. *Often plotting velocity vectors results in confusing mess. In these cases other flow visualizations are used like streamlines, streaklines, pathlines, and timelines. Find out what each represents and when they agree and differ.*

Exercise 66 (the energy equation). *Consider the energy equation for the temperature $T(x,t)$ of a fluid in motion.*

(a) *Review the derivation of the energy equation from section* 3 *of Chapter* 3 *when $-\Pi u$ is not negligible and the fluid is Newtonian. Simplify and nondimensionalize the energy equation.*

(b) *Under various simplifying assumptions, the energy equation can be written*

$$\rho_0 c_p \frac{D}{Dt} T(x,t) - \nabla \cdot (k \nabla T(x,t)) = 0.$$

Here c_p, k are the specific heat and thermal conductivity, respectively. Nondimensionalize this simplified energy equation.

Exercise 67 (Froude number). *Consider a flow in which buoyant forces due to gravity are significant. The model is simply the NSE plus a body force given by the gravitational force vector*

$$\nabla \cdot u = 0 \text{ and } \partial u/\partial t + u \cdot \text{ grad } u + \nabla(p/\rho_0) = (\mu/\rho_0)\Delta u + g.$$

Nondimensionalize this equation and show that two parameters occur:

$$\frac{\partial u^*}{\partial t^*} + u^* \cdot \nabla u^* = -\nabla p^* + \left(\frac{\mu}{\rho_0 \, VL}\right) \Delta^* u^* + \left(\frac{L}{V^2}\right) g.$$

The Froude number is thus

$$Fr := \frac{V}{\sqrt{L|g|}}.$$

The Froude number Fr *occurs when buoyant forces are significant. It measures the ratio of inertial forces to gravitational forces.*

Exercise 68 (vorticity boundary layers). *Consider the two-dimensional NSE. Take the curl of the equations (take $\frac{\partial}{\partial x}$ of the first momentum equation and $\frac{\partial}{\partial y}$ of the second, subtract, and use $\nabla \cdot u = 0$) to derive the two-dimensional vorticity equation. Repeat the scaling analysis to determine the scaling of ω in boundary layers.*

Exercise 69. *Whales are about* 100 *times larger than tunas and both swim at roughly the same speed. If* $\mathrm{Re}_{tuna} \sim 10^7$, *estimate* Re_{whales}. ***(Hint:*** *This estimate is easy.)*

Chapter 6

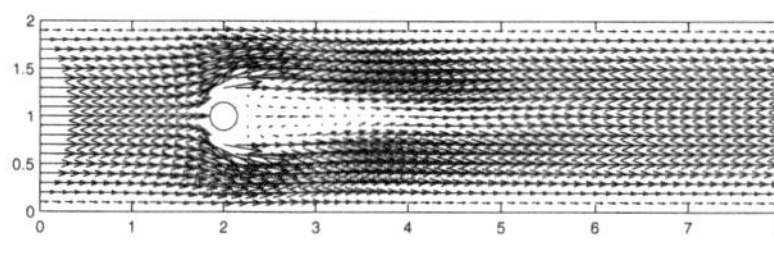

The Steady Navier–Stokes Equations

What is now proved was once only imagin'd.
W. Blake, *The Marriage of Heaven and Hell*, 1790.

6.1 The Steady Navier–Stokes Equations

Fluid particles refuse to move along straight lines....
Girard, in *Mémoires de la Classe des Sciences Mathématiques et Physiques de l'Institut de France*, 14, 1813.

Let us first understand the facts, and then we may seek for the causes.
Aristotle, *Nicomachean Ethics* (c. 325 BC).

Consider the problem of a steady (time-independent) flow in a domain Ω in 2 or 3 dimensions. The generic behavior of steady flows is a step up in complexity (and Reynolds number) beyond creeping flows. The classic experiment in fluid dynamics that illustrates this is the von Karman vortex street.

Example 16 (flow around a cylinder and the von Karman vortex street). Consider flow around a long cylinder which is uniform in the z-direction. This flow is typically visualized three ways: velocity vectors, streamlines, and contours of the vorticity.[49] In this flow, since the cylinder is uniform in one direction, if the incoming stream is uniform in that direction, the flow can often be modeled as two-dimensional. At low Reynolds numbers, the nonlinear terms in the NSE are negligible and the flow is very close to a solution of the Stokes problem (Figure 6.1). The streamlines are very simple (Figure 6.2).

At somewhat higher Reynolds numbers, the cylinder has a wake behind it (Figures 6.3 and 6.4). This is a nonlinear effect and it also increases the drag substantially, an effect called pressure drag.

[49]Recall that the vorticity is $\omega = \nabla \times u$. If the velocity u is a vector in $\mathbb{R}^2$, then the vorticity ω is a scalar.

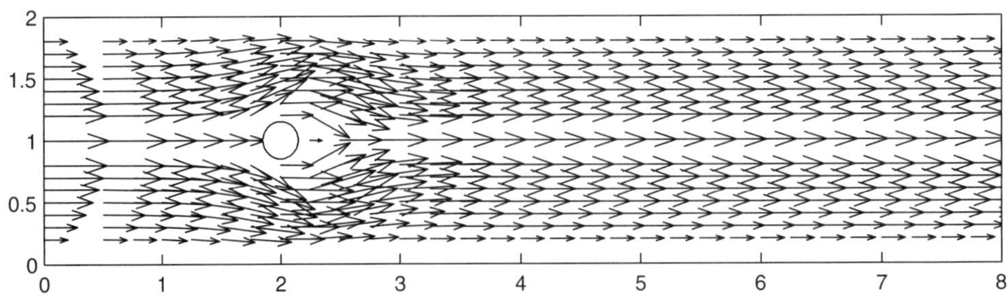

Figure 6.1. *Velocity vectors for* Re = 1.

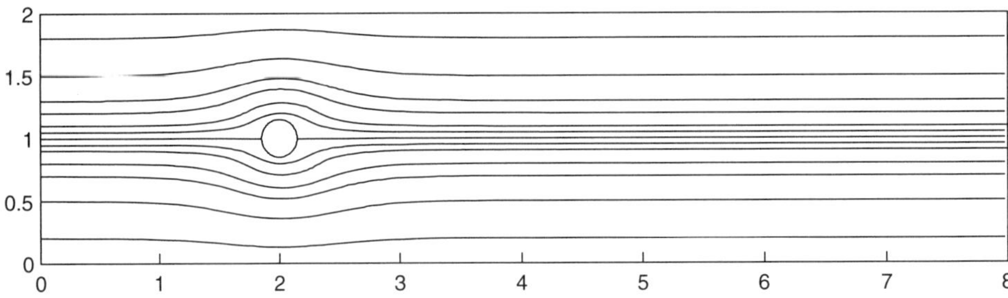

Figure 6.2. *Streamlines for* Re = 1.

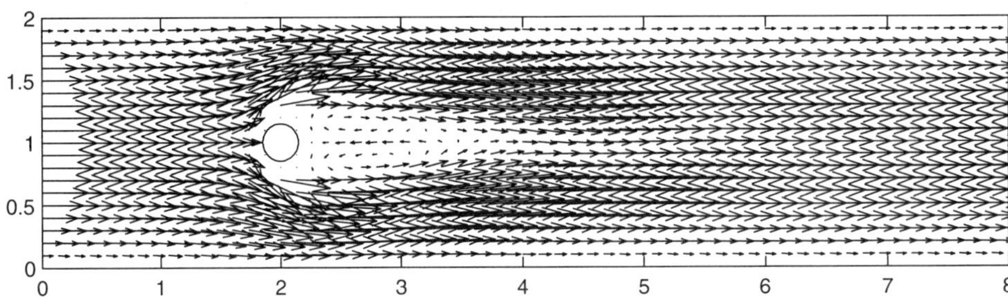

Figure 6.3. *Velocity vectors for* Re = 40.

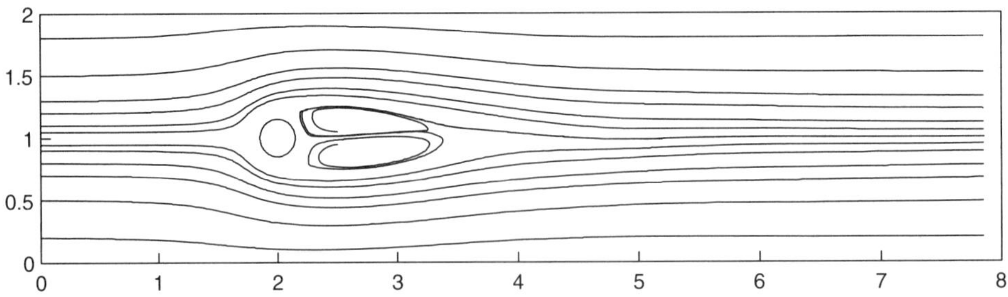

Figure 6.4. *Streamlines for* Re = 40.

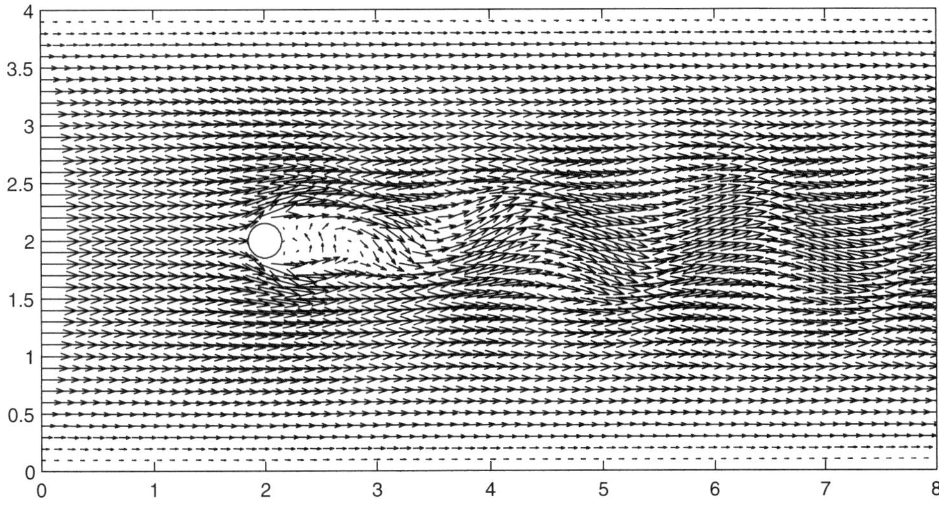

Figure 6.5. *Velocity vectors* Re = 200.

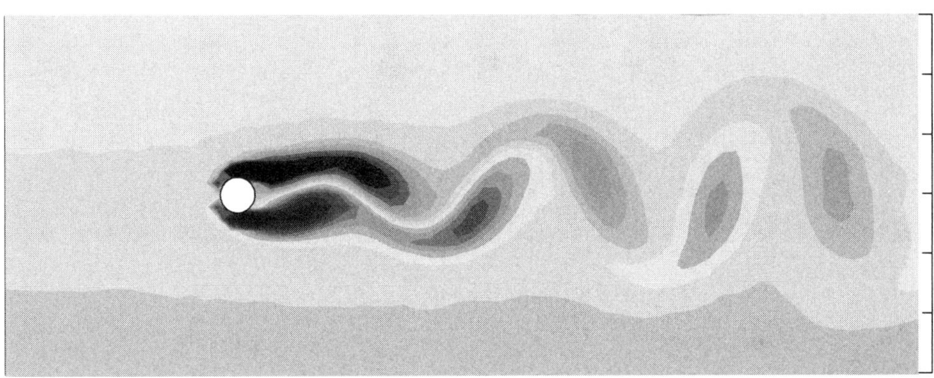

Figure 6.6. *Vorticity contours* Re = 200*: the von Karman vortex street.*

At somewhat higher Reynolds numbers, the eddy behind the cylinder becomes unstable. Then the steady solution is no longer observed in experiments because it becomes unstable[50] for the time-dependent problem. The eddies are then shed on alternate sides of the cylinder and the solution becomes periodic. This flow is known as the von Karman vortex street[51] (Figures 6.5 and 6.6).

It's interesting to see the pattern of the vortex street persists to quite a bit higher Reynolds number. We give plots at Re = 1000 and 2000 in Figures 6.7 and 6.8.

[50]Here *stable* means not observed in a physical experiment. This is connected to the mathematical condition that small perturbations cause the steady flow to become time-dependent and not return quickly to a steady flow.

[51]To illustrate the need for many stability concepts in fluid mechanics, the von Karman vortex street can be quite stable as a time-dependent solution of the time dependent NSE. As the figures in this remark show, vortex shedding can be computed quite easily by solving the time-dependent problem.

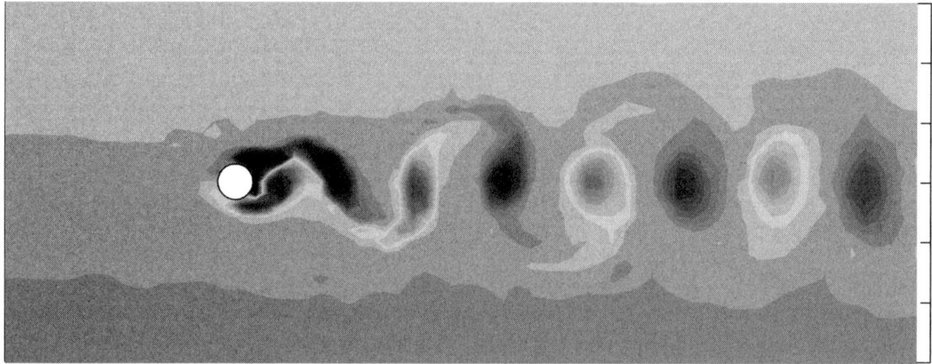

Figure 6.7. *Vorticity contours at* Re = 1000.

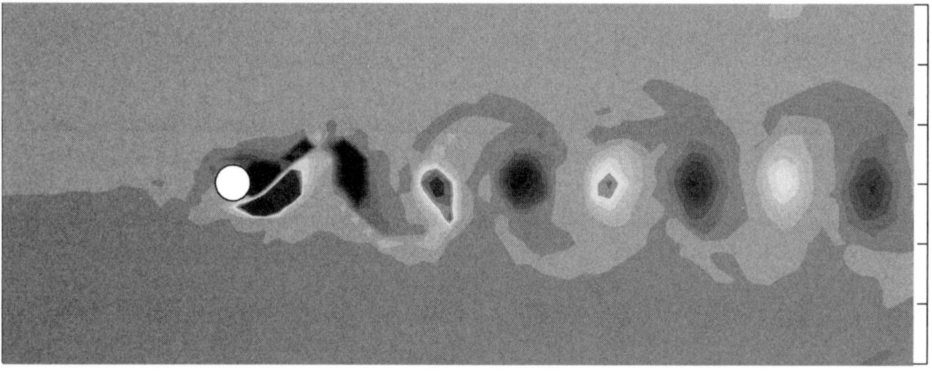

Figure 6.8. *Voticity contours at* Re = 2000.

If the mesh is generated adaptively, the mesh itself gives some insight into the flow. Figures 6.9 and 6.10 contain plots of the meshes for the Re = 40 and 200 simulations. ∎

The progression illustrated by the last example,

creeping flow → steady flow → time periodic flow → complex time dependent flow,

is observed for many different geometries, boundary conditions, and driving forces. Our study of steady flows will focus on internal flows which are important and which have a relatively simple mathematical formulation. Thus, consider a fluid contained by walls and driven by a body force f. The velocity and pressure (u, p) satisfy

$$\begin{aligned} &u \cdot \nabla u - \nu \Delta u + \nabla p = f \text{ and } \nabla \cdot u = 0 \text{ in } \Omega, \\ &u = 0 \text{ on } \partial\Omega. \end{aligned} \tag{6.1.1}$$

The pressure p is normalized to have mean value zero by $\int_\Omega p dx = 0$. The steady NSE are used to model flows in which nonlinear effects are important but are balanced by viscous

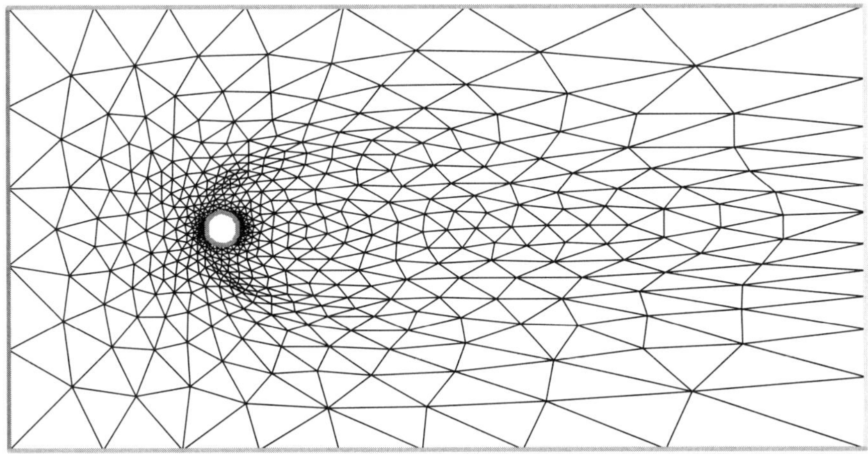

Figure 6.9. *FEM mesh for* Re = 40.

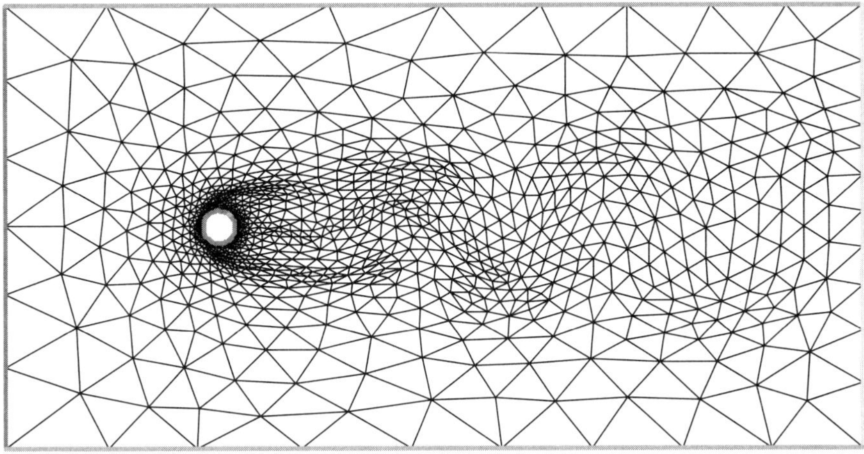

Figure 6.10. *FEM mesh for* Re = 200.

effects with the result that the noise in a real flow problem does not destabilize the motion and the flow will reestablish equilibrium. This chapter presents some of the mathematical theory of the steady NSE. This theory gives a careful and precise expression of the above physical picture. We prove that solutions are globally unique for small Reynolds number (interpreted correctly). Remarkably, for *any* Reynolds number, solutions of the steady NSE exist (section 3). These solutions can be locally unique and stable for the time-dependent problem or not. The flow can quickly return to steady state after small perturbations, or any small perturbation of the flow can result in a persistent time-dependent flow. Although steady solutions which are unstable in this latter sense cannot be realized in a physical experiment, they do influence the dynamics of the time-dependent problem. Of course, *not physically realizable* often is accompanied by *not computable by iterative methods* for solving the associated nonlinear and linearized problems, so our knowledge of steady solutions which are not stable for the time-dependent problem (described through an example in section 4) is sketchy at best.

We begin the study of steady flows by first studying their stability bounds and global energy balance, in other words, by finding the variational formulation of the steady NSE. Let $\Omega \subset \mathbb{R}^d (d = 2 \text{ or } 3)$ denote the region that encloses the fluid. We will seek the velocity and pressure (u, p) of the fluid motion in the region in the spaces (X, Q) given by

$$X := \{v : \Omega \to \mathbb{R}^d : v \in L^2(\Omega), \nabla v \in L^2(\Omega) \text{ and } v = 0 \text{ on } \partial\Omega\},$$
$$Q := \left\{q : \Omega \to \mathbb{R} : q \in L^2(\Omega) \text{ and } \int_\Omega q\,dx = 0\right\}.$$

Multiplying (6.1.1) by $v \in X$ and $q \in Q$ and integrating over Ω, the solution (u, p) satisfies

$$\int_\Omega u \cdot \nabla u \cdot v + \nu \nabla u : \nabla v - p \nabla \cdot v\,dx = \int_\Omega f \cdot v\,dx,$$

$$\int_\Omega q \nabla \cdot u\ dx = 0,$$

for all $v \in X, q \in Q$. We therefore convert (6.1.1) to the following weak form. Find $u \in X$, $p \in Q$ satisfying

$$(\nu \nabla u, \nabla v) + (u \cdot \nabla u, v) - (p, \nabla \cdot v) = (f, v)\ \forall v \in X,\ (q, \nabla \cdot u) = 0\ \forall q \in Q. \quad (6.1.2)$$

The form (6.1.2) writes the steady NSE as a pair of coupled equations. It is completely equivalent to writing it as one vector equation as follows. Find $(u, p) \in (X, Q)$ satisfying

$$(\nu \nabla u, \nabla v) + (u \cdot \nabla u, v) - (p, \nabla \cdot v) + (q, \nabla \cdot u) = (f, v)\ \forall (v, q) \in (X, Q).$$

Indeed, in the above picking alternately $(v, q) = (v, 0)$ and $(v, q) = (0, q)$ recovers the two equations in (6.1.2). Similarly, adding the two equations in (6.1.2) gives the above.

Recall that (X, Q) satisfies the continuous inf-sup condition (see Chapter 4) and the bilinear form $(q, \nabla \cdot v)$ is continuous on (X, Q). Two consequences are that the space of divergence free functions

$$V := \{v \in X : (\nabla \cdot v, q) = 0\ \forall q \in Q\}$$

is a well-defined, closed subspace of X and the weak formulations in (X, Q) and in V are equivalent. Hence V is a Hilbert space and (1.2) can be written equivalently as follows. Find $u \in V$ satisfying

$$(\nu \nabla u, \nabla v) + (u \cdot \nabla u, v) = (f, v)\ \forall v \in V. \quad (6.1.3)$$

Lemma 12 (skew-symmetry). *If $v, \nabla v \in L^2(\Omega)$, $\nabla \cdot v = 0$, and $v \cdot \hat{n} = 0$ on $\partial\Omega$*

$$\int_\Omega v \cdot \nabla v \cdot v\,dx = 0.$$

More generally,

$$\int_\Omega u \cdot \nabla v \cdot w \, dx = -\int_\Omega u \cdot \nabla w \cdot v \, dx$$

for any such u, v, w.

Proof. The first claim follows from the vector identity:

$$v \cdot \nabla v \cdot v = \frac{1}{2}\nabla \cdot (v|v|^2) - \frac{1}{2}|v|^2 \nabla \cdot v.$$

The second claim is left as an exercise to find the corresponding vector identity. Both identities are easy to derive using the subscript notation and summation convention. □

Using (6.1.3) and Lemma 12 gives an a priori bound on any solution of (6.1.3). Before doing so we need to recall the best norm for measuring the size of the body force f. Our strategy is the same as in the Stokes problem: set $v = u$ in the variational formulation and use some form of the Cauchy–Schwarz inequality on the right-hand side:

$$(f, u) \leq \{\text{some norm of } f\} \times ||\nabla u||.$$

Since $\nabla \cdot u = 0$, $u \in V$ so the best norm for this purpose is the dual norm, i.e., the norm on dual of V, recalled from Chapter 4 next.

Definition 24 (dual norms). *For* $f \in L^2(\Omega)$, *the* H^{-1} *norm and the* V^* *norm of* f *are, respectively,*

$$||f||_{-1} := \sup_{v \in X} \frac{(f, v)}{||\nabla v||}, \; ||f||_* := \sup_{v \in V} \frac{(f, v)}{||\nabla v||}.$$

The function spaces $H^{-1}(\Omega)$ *and* V^* *are, respectively, the closures of* $L^2(\Omega)$ *in, respectively,* $||\cdot||_{-1}$ *and* $||\cdot||_*$.

Operationally, the dual norm is the smallest norm satisfying

$$(f, v) \leq ||f||_* ||\nabla v|| \; \forall v \in V$$

and the H^{-1}*norm* is the smallest norm satisfying this inequality for any $v \in X$. Since $V \subset X \subset L^2(\Omega)$ and using the Poincaré–Friedrichs inequality, the spaces and norms are nested as

$$V \subset X \subset L^2(\Omega) \text{ and } ||f||_* \leq ||f||_{-1} \leq \frac{1}{C_{PF}}||f||.$$

Proposition 8 (stability of the velocity). *Any solution of* (6.1.3) *satisfies*

$$||\nabla u|| \leq \nu^{-1}||f||_*, \; \textit{where} \; ||f||_* = \sup_{v \in V} \frac{(f, v)}{||\nabla v||}.$$

Proof. Set $v = u$ in (6.1.3). Using $(u \cdot \nabla u, u) = 0$ gives[52]

$$\nu||\nabla u||^2 = (f, u) \leq ||f||_*||\nabla u||. \quad \square$$

6.2 Uniqueness for Small Data

I have often wished that mathematicians would devote themselves to improving the rigor of this and similar demonstrations, rather than to inventing special methods for the particular cases.

Lord Rayleigh, in a letter to Maxime Bôcher, quoted in Birkhoff, *Hydrodynamics*, Princeton, 1950.

We consider the question, Under what conditions can we guarantee there is *at most* one solution to (6.1.3)? We begin with an important technical lemma.

Lemma 13 (continuity of the trilinear form). *There is a finite constant $M = M(\Omega)$ such that for all $u, v, w \in X$,*

$$|(u \cdot \nabla v, w)| \leq M||\nabla u||\ ||\nabla v||\ ||\nabla w||. \tag{6.2.1}$$

Proof. Recall that Hölder's inequality implies

$$\left|\int_\Omega f\ g\ h\ dx\right| \leq ||f||_{L^p}||g||_{L^q}||h||_{L^r}, \quad \frac{1}{p} + \frac{1}{q} + \frac{1}{r} = 1,$$

where $1 \leq p, q, r \leq \infty$.

Using this,

$$|(u \cdot \nabla v, w)| \leq ||u||_{L^p}||\nabla v||\ ||w||_{L^r}, \quad \frac{1}{p} + \frac{1}{2} + \frac{1}{r} = 1.$$

Now, by the Ladyzhenskaya inequalities and the Poincaré–Friedrichs inequality (see Chapter 1), $||u||_{L^4} \leq C||\nabla u||$ in two and three dimensions. The result now follows by picking $p = r = 4$. $\square$

What is the best possible constant M in this lemma? Obviously, it is

$$M := \sup_{u,v,w \in X} \frac{(u \cdot \nabla v, w)}{||\nabla u||||\nabla v||||\nabla w||},$$

which is finite by Lemma 13. If an estimate of the form of (6.2.1) is needed where $u, v, w \in V$, i.e., are div-free, then the best constant is the famous $N = N(\Omega)$ defined by

$$N := \sup_{u,v,w \in V} \frac{(u \cdot \nabla v, w)}{||\nabla u||||\nabla v||||\nabla w||}. \tag{6.2.2}$$

[52]It is a good thing that $(u \cdot \nabla u, u) = 0$! Otherwise there is almost no hope for a bound holding *in the large*, i.e., for general velocities. This is because the bad term $(u \cdot \nabla u, u)$ is cubic in the velocity, while the good term (the positive definite viscous term) is quadratic in the velocity. There is no hope in showing that a cubic term is dominated by a quadratic term for large velocities (if $|u| \sim 10$, compare 10^3 to 10^2). On the other hand, if the boundary conditions are such that $(u \cdot \nabla u, u)$ is nonzero, generally, a theory *in the small* meaning for small forces and velocities is possible (if $|u| \sim .1$, compare $.1^3$ to $.1^2$).

Since $V \subset X$ and since taking supremums over larger sets makes the supremum increase, we have

$$0 < N \leq M < \infty.$$

With these preliminaries, we can now give the following global uniqueness result.

Proposition 9 (uniqueness for small data). *Suppose the data satisfies the smallness condition:*

$$M\nu^{-2}||f||_* < 1. \tag{6.2.3}$$

Then, (6.1.3) *can have at most one solution. The same conclusion holds if*

$$N\nu^{-2}||f||_* < 1. \tag{6.2.4}$$

Proof. There is one basic proof of uniqueness: assume not, subtract, and show the difference is zero. Indeed, let u_1, u_2 be two solutions of (6.1.3). Setting $\phi = u_1 - u_2$ and subtracting gives

$$(\nu\nabla(u_1 - u_2), \nabla v) + (u_1 \cdot \nabla u_1, v) - (u_2 \cdot \nabla u_2, v) = 0 \; \forall v \in V.$$

Setting $v = \phi = u_1 - u_2$ gives

$$\nu||\nabla\phi||^2 + (u_1 \cdot \nabla(u_1 - u_2) + (u_1 - u_2) \cdot \nabla u_2, \phi) = 0,$$

or, using $(u_1 \cdot \nabla\phi, \phi) = 0$,

$$\nu||\nabla\phi||^2 + (\phi \cdot \nabla u_2, \phi) = 0.$$

Using Lemma 13 gives

$$\nu||\nabla\phi||^2 \leq M||\nabla\phi||^2||\nabla u_2|| \leq (\text{using the bound on } ||\nabla u_2|| \,) \leq M\, \nu^{-1}||f||_*||\nabla\phi||^2.$$

Thus, $(1 - M\, \nu^{-2}||f||_*)||\nabla\phi||^2 \leq 0$, proving the theorem. □

Remark 5 (more on uniqueness). *The equation* $\nu||\nabla\phi||^2 + (\phi \cdot \nabla u, \phi) = 0$ *bears careful consideration. (It might contain the key to winning the million-dollar Clay prize.) Writing it out gives*

$$\nu||\nabla\phi||^2 = -(\phi \cdot \nabla u, \phi) = -\int_\Omega \phi_i\phi_j u_{j,i} dx.$$

The tensor $\phi_i\phi_j$ *is obviously symmetric and the integrand is the contraction of this with* ∇u*. Because it is symmetric, only the contraction with the symmetric part of* ∇u *is nonzero:*

$$\phi_i\phi_j u_{j,i} = (\phi_i\phi_j) : (u_{j,i}) = \phi_i\phi_j \frac{u_{j,i} + u_{i,j}}{2}$$

or

$$\nu||\nabla\phi||^2 = \int_\Omega \phi \cdot (-\nabla^s u) \cdot \phi dx.$$

The trace of the deformation tensor $\nabla^s u$ *is* $\nabla \cdot u = 0$ *. Thus* $\nabla^s u$ *has both positive and negative eigenvalues that sum to zero. From the above, the positive eigenvalues of* $\nabla^s u$ *actually help prove uniqueness, while viscosity actually only needs to battle the negative* *one's effects. Thus, loosely speaking, one central and huge open problem is to quantify when and how in the large/integrated over Ω the effects of the positive, good eigenvalues outweigh those of the bad, negative ones. At this point, we are being deliberately fuzzy about what "outweigh" really means (because it is not known as well as one would hope). This will be somewhat clearer after studying the time-dependent problem. As soon as we apply an inequality like those in Lemmas* 12 *and* 13, *we get no help from the good eigenvalues and thus are assuming the the worst that can happen always does happen.*

In some cases, other bounds on the nonlinear term can be useful or even essential. Another useful continuity estimate is presented next.

Lemma 14. *For $u, v, w \in X$,*

$$|(u \cdot \nabla v, w)| \leq C(\Omega)||u||^{\frac{1}{2}}||\nabla u||^{\frac{1}{2}}||\nabla v||||\nabla w|| \text{ for } d = 2, 3 \text{ and}$$
$$|(u \cdot \nabla v, w)| \leq C(\Omega)||u||^{\frac{1}{2}}||\nabla u||^{\frac{1}{2}}||\nabla v||||w||^{\frac{1}{2}}||\nabla w||^{\frac{1}{2}} \text{ for } d = 2.$$

Proof. This is an exercise in applying Hölder's inequality and the Sobolev imbedding theorem. □

Anytime there is a *global uniqueness* result you should suspect that somewhere in the background a global contraction map is lurking. If it is there, it is worth finding it since it gives also an existence result and a convergent iterative process. To find the contraction map behind Proposition 10 we need to introduce the Oseen problem.[53]

6.2.1 The Oseen problem

Given a vector field $\vec{b}$ which is div-free, vanishes on $\partial\Omega$, and is smooth enough, the Oseen problem is to find (u, p) satisfying

$$-\nu\Delta u + \vec{b} \cdot \nabla u + \nabla p = f, \ \nabla \cdot u = 0 \text{ in } \Omega,$$
$$u = 0 \text{ on } \partial\Omega \ \left(\text{and } \int_\Omega p\, dx = 0\right).$$

The Oseen problem has the variational formulation: find $u \in V$ satisfying

$$a(u, v) := \nu(\nabla u, \nabla v) + (\vec{b} \cdot \nabla u, v) = (f, v) \ \forall v \in V. \tag{6.2.5}$$

Lemma 15. *Let $\vec{b} \in V$ be given. Then $a(\cdot, \cdot)$ is continuous and coercive:*

$$a(u, u) \geq \nu||\nabla u||^2 \ \forall\, u \in V,$$
$$a(u, v) \leq (\nu + M(\Omega)||\nabla b||)||\nabla u||\ ||\nabla v|| \ \forall u, v \in V.$$

Proof. This follows from the Cauchy–Schwarz inequality and Lemma 13. □

[53] C. W. Oseen (1879–1944) was one of the mathematical pioneers of the Navier–Stokes equations.

Corollary 2 (solvability of the Oseen problem). *Consider $f \in X^*$ to be fixed. For any $\vec{b} \in V$ the Oseen problem has a unique solution $u \in V$. The solution u satisfies*

$$\nu||\nabla u|| \leq ||f||_*.$$

Proof. This follows from the Lax–Milgram theorem. $\square$

Consider the map which takes the velocity field $\vec{b}$ to the solution u of (6.2.5). Let this be denoted by

$$u = T(b) \quad \text{so} \quad T : V \to V.$$

If $u^* \in V$ is a fixed point of T, $u^* = T(u^*)$, then u^* satisfies (6.2.5) with $\vec{b}$ replaced by u^*—exactly the steady NSE!

Proposition 10. *$T : V \to V$ is uniformly bounded: if $u = T(b)$, then*

$$||\nabla u|| \leq C(\nu, \Omega, ||f||_*) < \infty \text{ (i.e., independent of } ||\nabla \vec{b}\,||).$$

Further, if (6.2.3) *(or* (6.2.4)*) holds, T is a global contraction.*

Proof. Exercise! $\square$

Corollary 3 (existence for small data). *Under* (6.2.3) *(or* (6.2.4)*) a solution to the steady NSE exists.*

Instead of proving existence by applying an abstract fixed point theorem, we will prove that the following associated iteration is globally convergent. Given $u^0 \in V$ calculate $u^1, u^2, \cdots \in V$ satisfying

$$\nu(\nabla u^{n+1}, \nabla v) + (u^n \cdot \nabla u^{n+1}, v) = (f, v) \; \forall v \in V.$$

Proposition 11. *Under* (6.2.3) *(or* (6.2.4)*) the above iterates u^n form a Cauchy sequence and thus converge, $u^n \to u$ as $n \to \infty$, where u is the unique solution of the NSE.*

Proof. We prove $u^n \to u$ and leave it as an exercise to prove the remainder.[54] Rewrite (6.2.3) as

$$M\nu^{-2}||f||_* \leq \alpha < 1.$$

Call $e^n = (u - u^n) \in V$. Subtraction gives

$$\nu(\nabla e^{n+1}, \nabla v) + (u \cdot \nabla u - u^n \cdot \nabla u^{n+1}, v) = 0.$$

Notice the pattern $ab - cd$ in the quadratic nonlinearity. This is often treated by

$$ab - cd = (ab - bc) + (bc - cd) = b(a - c) + c(b - d).$$

[54]Try the proof! It is simplest to break it into three steps. For the first, prove the observation that if $||\nabla(u^{n+1} - u^n)|| \leq \alpha||\nabla(u^n - u^{n-1})||$ holds for all n for some $\alpha < 1$, then the sequence is Cauchy. (This is the key and more generally useful.) Second, prove the uniform bound $\nu||\nabla u^n|| \leq ||f||_*$ for all n. Third, mimic the convergence proof to complete the proof.

Setting $v = e^{n+1}$ and adding and subtracting $u^n \cdot \nabla u$ gives

$$\nu||\nabla e^{n+1}|| = -[(e^n \cdot \nabla u, e^{n+1}) - (u^n \cdot \nabla e^{n+1}, e^{n+1})] = -(e^n \cdot \nabla u, e^{n+1}),$$

by skew symmetry (Lemma 12). Thus, since $||\nabla u|| \leq \nu^{-1}||f||_*$,

$$\nu||\nabla e^{n+1}||^2 \leq M||\nabla e^n||\,||\nabla u||\,||\nabla e^{n+1}|| \leq \left(\frac{M}{\nu}||f||_*\right)||\nabla e^n||\,||\nabla e^{n+1}||,$$

or $||\nabla e^{n+1}|| \leq \alpha||\nabla e^n||$ and $||\nabla e^n|| \to 0$ as $n \to \infty$. □

6.3 Existence of Steady Solutions

> *Now,* existence *means one thing to the physicist, quite another to the mathematician. Whether a thing exists or does not exist is to the physicist a matter of experimental evidence; to the mathematician, on the other hand, existence is coextensive with freedom from contradiction.*
>
> T. Dantzig, in [21, p. 38].

The steady NSE have solutions for any Reynolds number and for quite general regular domains and quite general body forces $f \in L^2(\Omega)$. There are special flow problems where this can be seen by checking that an explicit formula for the velocity and pressure is always a solution.

Example 17. Consider flow between parallel plates (Figure 6.11). Let the flow domain be the unit cube $0 \leq x, y, z \leq 1$. On the left, right, front, and back boundaries, impose periodic boundary conditions. On the bottom boundary impose $u = 0$ for $z = 0$. The top boundary moves to the right with constant velocity U so the no-slip boundary condition at the top is $u = (0, U, 0)$. This flow can also be represented in two dimensions as one slice of the three-dimensional flow that is uniform in one direction. This flow is called a shear flow. It is driven by the motion of the top wall so its body force is $f = 0$. It is a simple exercise to check that

$$u = (0, Uz, 0), 0 \leq z \leq 1, p = C,$$

is a solution of the steady NSE for every Reynolds number.[55]

Thus a solution to this flow problem exists for any Re. For higher velocities, it is no longer observed in physical experiments but it still exists as a solution of the steady NSE. ∎

Any solution of the steady NSE

$$-\nu\Delta u + u \cdot \nabla u + \nabla p = f,\ \nabla \cdot u = 0 \text{ in } \Omega, \quad (6.3.1)$$
$$u = 0 \text{ on } \partial\Omega \left(\text{and } \int_\Omega p\, dx = 0\right)$$

[55] This exact solution shares a property of most of the few known exact solutions of the NSE: it is an exact solution of the *Stokes problem* for which $u \cdot \nabla u \equiv 0$.

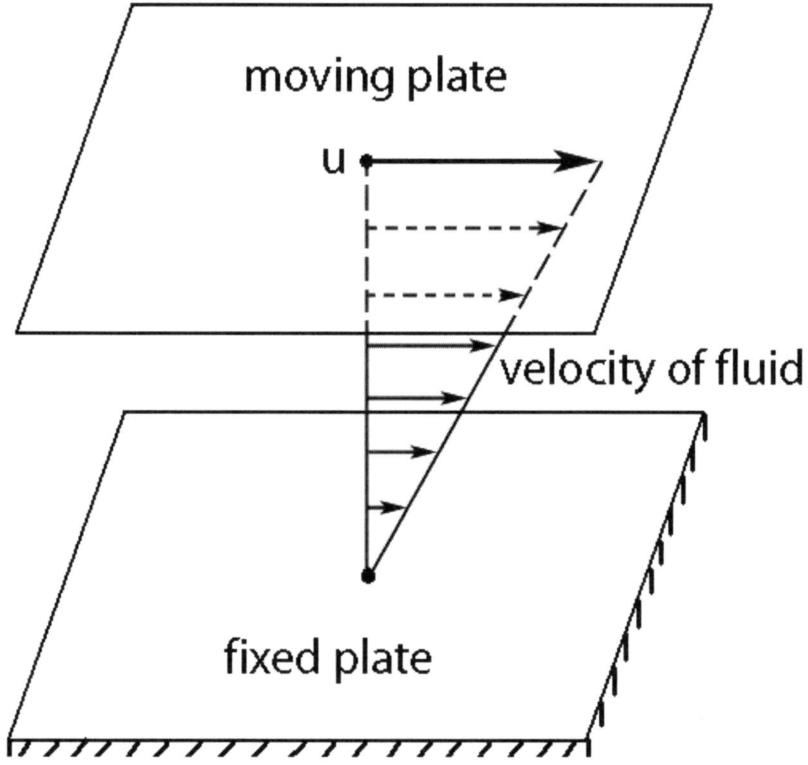

Figure 6.11. *Shear flow between parallel plates.*

must satisfy the a priori bound

$$||\nabla u|| \leq \frac{1}{\nu}||f||_*.$$

Such an a priori bound is *almost* enough by itself to ensure that solutions to (6.3.1) exist. The proof of existence, using this fact, is based on the Leray–Schauder fixed-point theorem.

Definition 25 (compact sets and operators). *A set G in a Hilbert space Y is compact if every sequence in G has a convergent subsequence whose limit is in G.*

An operator $\mathcal{F}:Y \to Y$ is compact if it maps bounded sets to sets with compact closure.

Compactness is often verified by using the Reillich lemma, which states that the imbedding

$$H^k(\Omega) \hookrightarrow H^j(\Omega)$$

is compact for $k > j$. For example, a sequence $u_n \in X$ with

$$||\nabla\nabla u_n|| \leq K \ \forall\, n$$

must have a convergent subsequence in X: there is a u_{n_j} and $u \in X$ with $u_{n_j} \to u$ in X as $n_j \to \infty$. Also, an operator that is smoothing will generally be compact.

The existence proof in the steady case hinges on another fundamental contribution of Leray—the Leray–Schauder fixed-point theorem.

Theorem 16 (the Leray–Schauder fixed-point theorem). *Let Y be a Hilbert space and let $\mathcal{F} : Y \to Y$ be a compact map. Consider the fixed point problem: find $y^* \in Y$ such that*

$$y^* = \mathcal{F}(y^*). \tag{6.3.2}$$

Associate with (6.3.2) *the family of fixed point problems: find $y_\lambda \in Y$ such that*

$$y_\lambda = \lambda \mathcal{F}(y_\lambda), \quad 0 \le \lambda \le 1. \tag{6.3.3}$$

If there is a constant K such that all solutions of (6.3.3) *are uniformly bounded*

$$||y_\lambda||_Y \le K \quad \text{for } 0 \le \lambda \le 1,$$

then there exists a solution y^* to (6.3.2).

The plan of attack is now simple: reformulate (6.3.1) as a fixed-point problem, insert a parameter, and adapt the proof of the a priori bound to give a bound uniform in λ. To this end, let $T : X^* \to V$ be the solution operator of the Stokes problem. Specifically, $T(g) = u$, where $u \in V$ solves

$$-\nu(\nabla u, \nabla v) = (g, v) \; \forall \, v \in V. \tag{6.3.4}$$

The Lax–Milgram theorem ensures T exists and is bounded

$$||T|| = \sup_{g \in X^*} \frac{||T(g)||_X}{||g||_*} = \sup_{g \in X^*} \frac{||\nabla u||}{||g||} \le \frac{1}{\nu} \text{ since } ||\nabla u|| \le \frac{1}{\nu}||g||.$$

Further, it is easy to check that T is linear. Define the nonlinear operator

$$N(u) = f - u \cdot \nabla u.$$

Lemma 16. $N : V \to X^*$ *is a continuous and bounded operator.*

Proof. First we shall verify that N is bounded (maps bounded sets to bounded sets). For this we must show

$$||f - u \cdot \nabla u||_* \le C(||\nabla u||) < \infty.$$

Now,

$$||f - u \cdot \nabla u||_* \le ||f||_* + ||u \cdot \nabla u||_*.$$

Thus,

$$||u \cdot \nabla u||_* = \sup_{v \in X} \frac{(u \cdot \nabla u, v)}{||\nabla v||} \le M(\Omega)||\nabla u||^2 = C(||\nabla u||) < \infty$$

and $N(u)$ is bounded.

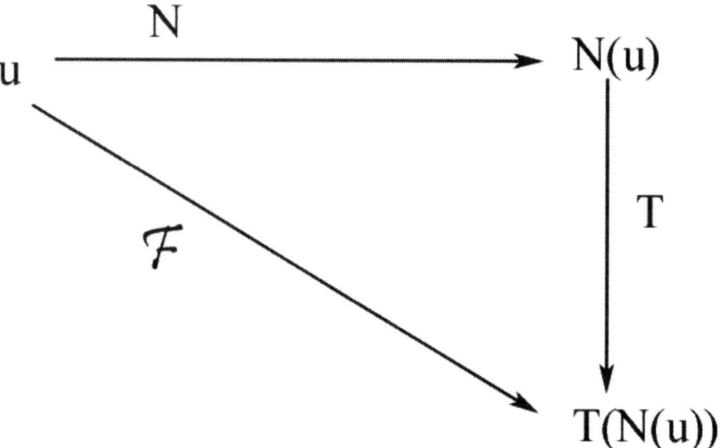

Figure 6.12. *The map T(N(u)).*

We leave it as an interesting exercise to show $||N(u_n) - N(u)||_* \to 0$ if $||\nabla(u_n - u)|| \to 0$. ◻

Lemma 17. *Define*

$$\mathcal{F} : V \to V \text{ by } \mathcal{F}(u) = T(N(u)).$$

(i) $\mathcal{F} : V \to V$ *compactly.*

(ii) u *is a solution of the steady NSE* (6.1.3) *if and only if* u *is fixed point of* $\mathcal{F}$.

Proof. (i) We shall prove only that $\mathcal{F} : V \to V$ continuously.[56] Since $T : V^* \to V$ all we need to verify is that $N : V \to V^*$ continuously (see Figure 6.12).

Exactly this was verified in the previous lemma; thus (i) is proven.

For (ii), replace g by $f - u \cdot \nabla u$. This shows a fixed point of $\mathcal{F}$ satisfies: $u \in V$ and

$$\nu(\nabla u, \nabla v) = (f, v) - (u \cdot \nabla u, v) \ \forall \ v \in V,$$

which is exactly the steady NSE. ◻

Theorem 17 (existence for all data). *For any* ν *and any* $f \in X^*$, *a solution* u *to the steady NSE exists.*

Proof. We shall show that a fixed point u of

$$u = \mathcal{F}(u) \quad \text{in } V$$

[56]The proof of compactness is beyond, but not very far beyond, the mathematics developed in Chapters 1 through 6. Briefly, using the Ladyzhenskaya inequalities and the Sobolev embedding theorem, the bound on the nonlinear term can be sharpened using fractional order Sobolev spaces to read $||N(u)||_{V^*} \leq C||u||^2_{H^{3/4}}$. The embeding $H^1 \hookrightarrow H^{3/4}$ is compact. So $N : V \hookrightarrow H^{3/4} \to V^*$ is continuous after compact and thus is compact.

exists. Consider

$$u_\lambda = \lambda \mathcal{F}(u_\lambda) \text{ in } V.$$

By the Leray–Schauder fixed point theorem we need only prove an á priori bound on $||\nabla u_\lambda||$ independent of λ. Unscrambling the u_λ problem,

$$u_\lambda = T(\lambda f - \lambda u_\lambda \cdot \nabla u_\lambda),$$

which holds if and only if $u_\lambda \in V$ satisfies

$$\nu(\nabla u_\lambda, \nabla v) = (\lambda f, v) - (\lambda u_\lambda \cdot \nabla u_\lambda, v) \ \forall \ v \in V.$$

Setting $v = u_\lambda$ shows

$$\nu||\nabla u_\lambda|| \leq \lambda||f||_*.$$

Thus, for $0 \leq \lambda \leq 1$,

$$||\nabla u_\lambda|| \leq \frac{\lambda}{\nu}||f||_* \leq \frac{1}{\nu}||f||_* =: K,$$

and a solution exists. □

Remark 6. *It is interesting to note that solutions always exist for the steady problem, even when they are not observed in nature! These steady solutions which exist, but are not observed, are solutions of the steady problem that are unstable when considered as solutions of the time-dependent problem. (This type of physical stability is one of many important stability concepts in fluid mechanics.) As such, they are still very interesting since both stable and unstable steady solutions play a large role in determining the dynamic behavior of the time-dependent problem.*

Discretizations of the steady NSE are nonlinear and must be solved by some iterative process (like Newton's method). If we think of the iteration number as a time-step number, often iteration is equivalent to time-stepping to steady state. Thus it can be very hard to compute physically unstable, steady state solutions. On the other hand, if the time-dependent NSE are solved, it can be quite easy to compute time-dependent solutions like the von Karman vortex street.

6.4 The Structure of Steady Solutions

> *If the Reynolds number* Re *is small, then the Navier–Stokes equations have a unique solution which can be observed in nature. When* Re *is large, even when a solution can be obtained, it is not observed in nature. The reason lies in the fact that the solutions are unstable; very small perturbations, too small to be measured by the experimenter, can be amplified and induce large changes in the flow.*
>
> A. J. Chorin, in *Lectures on Turbulence Theory*, Publish or Perish, Boston, 1975.

The description given earlier of the Taylor experiment gives an interesting window into the behavior of steady solutions. Briefly,

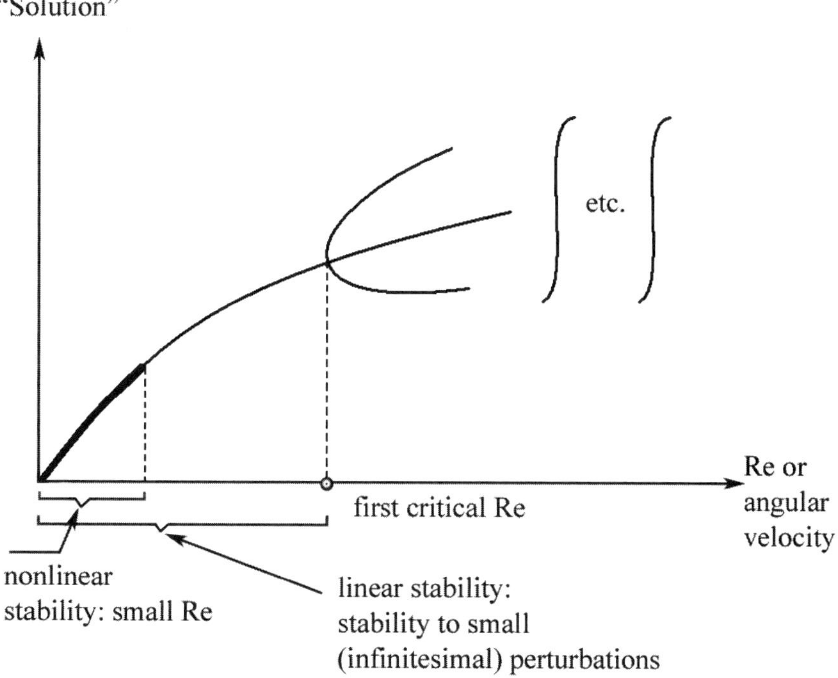

Figure 6.13. *Behavior of flow between rotating cylinders.*

- For very small Re, there is a steady solution (known explicitly) where each fluid part simply revolves around the cylinder's axis of rotation. This closed-form velocity is an exact solution for all Re.
- Nevertheless, as Re increases to a critical value the flow abruptly reorganizes into Taylor cells. There are thus three solutions at these Re's the observed Taylor cells, the same with rolling motion reversed, and the original (now unstable) rotational motion.
- At a higher still critical Re, the number of Taylor cells abruptly increases so one can observe five solutions.
- This continues until a quasi-periodic, time-dependent motion is observed ("wavy" cells).
- At still higher Re, these break down into turbulent bands and then fully turbulent flow.

This early behavior is often depicted by the schematic in the Figure 6.13.This schematic suggests that solutions to the steady NSE are typically not unique and, except for a few critical Reynolds numbers, separated from one another in V. We shall see (without proofs[57]) that this picture is remarkably accurate.

[57]For clear proofs, see, e.g., [19], [95].

Definition 26 (isolated solution of the NSE). *$u \in V$ is an isolated solution of the NSE if there is a positive radius $r > 0$ such that any other solution $w \in V$ is at least distance r from u:*

$$||\nabla(u - w)|| \geq r.$$

The above schematic suggests that, except at a few critical Re, all solutions are isolated. Actually, more can be said. Given $u^* \in V$, the linearized (at u^*) Navier–Stokes system is as follows: given $f \in V^*$, find (w, q) satisfying

$$\begin{aligned} &-\nu\Delta w + u^* \cdot \nabla w + w \cdot \nabla u^* + \nabla q = f \text{ in } \Omega, \\ &\nabla \cdot w = 0 \text{ in } \Omega, \\ &w = 0 \text{ on } \partial\Omega \text{ (and } \textstyle\int_\Omega q\, dx = 0). \end{aligned} \tag{6.4.1}$$

This is equivalent to the following: for $u^* \in V$, given $f \in V^*$ find $w \in V$ satisfying

$$\nu(\nabla w, \nabla v) + (u^* \cdot \nabla w, v) + (w \cdot \nabla u^*, v) = (f, v)\ \forall\, v \in V. \tag{6.4.2}$$

Definition 27 (nonsingular solution of the NSE). *$u^* \in V$ is a* nonsingular *solution of the steady NSE if*

(i) *u^* is a solution of the steady NSE,*

(ii) *the linearized problem* (6.4.2) *has a unique solution for any $f \in V^*$, and*

(iii) *the map $T(u^*) : V^* \to V$ by $T(u^*)f = w$, the solution of* (6.4.2) *is a bounded linear operator.*

There are several mathematical reasons why nonsingular solutions are natural in some sense. The first is that they are isolated and satisfy a condition which implies a stability bound.

Theorem 18. *Let $u^* \in V$ be a nonsingular solution of the steady NSE. Then,*

(i) *u^* is an isolated solution, and*

(ii) *u^* satisfies the following inf-sup nonsingularity condition: there is $\delta = \delta(\nu, \Omega, u^*) > 0$ such that*

$$\inf_{0 \neq w \in V} \sup_{0 \neq v \in V} \frac{\nu(\nabla w, \nabla v) + (u^* \cdot \nabla w, v) + (w \cdot \nabla u^*, v)}{||\nabla w||\, ||\nabla v||} \geq \delta > 0,$$

or, equivalently, for any $w \in V$,

$$\sup_{0 \neq v \in V} \frac{\nu(\nabla w, \nabla v) + (u^* \cdot \nabla w, v) + (w \cdot \nabla u^*, v)}{||\nabla v||} \geq \delta\, ||\nabla w||. \tag{6.4.3}$$

Proof. The first is essentially an application of the mean value theorem. The second part is equivalent to the solution operator $T(u^*)$ to (6.4.2) existing. See, e.g., Girault and Raviart [45] for a complete proof. □

The next theorem asserts that solution to the steady NSE is generically nonsingular. This result shows that the *velocity* inf-sup condition (6.4.3) for boundedness of the velocity must be the key assumption on the solution of the steady NSE (at moderate Reynolds numbers) upon which its numerical analysis must be based.

Theorem 19 (generic nonsingularity). *For any* $\mathrm{Re}^{-1} > 0$*, there is an open dense set* $\mathcal{O} = \mathcal{O}(\mathrm{Re}) \subset L^2(\Omega)$ *such that for any* $f \in \mathcal{O}$*, the set of solutions of the steady NSE is finite and odd in number. Each solution for these* $f \in \mathcal{O}$ *is nonsingular. Further, on each connected component of* $\mathcal{O}$ *the number of solutions is constant and each solution depends smoothly on* f.

6.5 Remarks on Chapter 6

The best place to begin further study of the steady NSE is in Chapter 8 of [38] and [45]. The picture of the road to turbulence beginning with period doubling is described in [54] and [30]. The theorem on generic nonsingularity of solutions to the steady NSE is explained well in, for example, [19] and [95].

The sharpened bounds on the nonlinear term depend on a simple application of Hölder's inequality and a less simple use of the Sobolev embedding theorem and interpolation inequalities. The following are two examples, useful in the theory and error analysis of NSE, of the type of inequalities that can result. For all $v \in X$,

$$\begin{aligned} ||v||_{L^4(\Omega)} &\leq C||u||^{\frac{1}{2}}||\nabla u||^{\frac{1}{2}} \text{ if } \Omega \subset \mathbb{R}^2, \\ ||v||_{L^4(\Omega)} &\leq C||u||^{\frac{1}{4}}||\nabla u||^{\frac{3}{4}} \text{ if } \Omega \subset \mathbb{R}^3, \text{ and} \\ ||v||_{L^6(\Omega)} &\leq C||\nabla u|| \text{ if } \Omega \subset \mathbb{R}^3. \end{aligned} \tag{6.5.1}$$

For equilibrium flow problems driven by body forces and not by some inflow velocity, specifying the characteristic velocity in the Reynolds number is unclear. Often the complexity of such problems is characterized in terms of the Grashof number instead of the Reynolds number. The Grashof number is given by

$$Gr := \frac{L^{3-\frac{d}{2}}}{\rho_0 \nu^3}||\text{body} - \text{force}||,$$

where d denotes the dimension of the flow domain Ω (so $d = 2$ or 3) and $L = \text{diameter}(\Omega)$. The body force is the original body force before rescaling by density (as we have done). If the body force $f \in L^2(\Omega)$, the small data condition (6.2.3) can be written in terms of the Grashof number. First the scaling of the constant $M(\Omega)$ and the relative scalings with respect to L of the L^2 norm and V^* norm with respect to L must be determined (in the usual way). After these steps the result is that $M\nu^{-2}||f||_* = C \cdot Gr$ so that *small data* means exactly *small Grashof number*.

6.6 Exercises

Exercise 70 (the pressure). *Begin with Proposition* 8. *Use the inf-sup condition and* (6.1.2) *to prove an á priori bound on the pressure p.* (***Hint:*** *This follows the case of creeping flow closely.*)

Exercise 71. *Why is a golf ball dimpled? This is a simple question about flow over an obstacle which, like many such questions in fluid dynamics, has a nonsimple answer. Do some reading and explain why dimpling reduces drag on golf balls but would increase drag on cars, for example.*

Exercise 72. *Consider the proof of Proposition* 11. *First complete the proof. Next, rework the proof using Lemma* 14 *instead of Lemma* 13. *Does the result change?*

Exercise 73. (a) *Prove that solutions are globally unique provided*

$$N\,\nu^{-2}||f||_* < 1$$

(i.e., finish the proof of the global uniqueness proposition).

(b) *The constants N and M depend on the geometry of and the size of Ω. Pick N or M and determine its dependence on $L = \mathrm{diam}(\Omega)$.* (***Hint:*** *Change variables $x \leftarrow x/L$.)*

Exercise 74. *Consider the nonlinear elliptic BVP in $\Omega \subset \mathbb{R}^2$. Given $f \in L^2(\Omega)$, find u satisfying*

$$-\kappa\Delta u + |u|u = f \text{ in } \Omega,\ u = 0 \text{ on } \partial\Omega.$$

Find the variational formulation of this problem and prove an á priori bound on its solution.

Prove solutions are unique under a small data condition. Prove solutions exist under this same condition.

Exercise 75. *Let $N(v) := f - v \cdot \nabla v$. Consider $F : X \to X$ by $F(v) = T(N(v))$. Under a global uniqueness condition, show that F is a contraction.*

Exercise 76. *For $N(u) = f - u \cdot \nabla u$, show $N : X \to X^*$ continuously.*

Exercise 77. *Let $Y = \mathbb{R}^n$ (for n finite). Show that* any *continuous function $F : \mathbb{R}^n \to \mathbb{R}^n$ is compact.*

Exercise 78. *Consider the square linear system $A_{n\times n}x_{n\times 1} = b_{n\times 1}$. Let $\mathcal{F} : \mathbb{R}^n \to \mathbb{R}^n$ be $\mathcal{F}(x) = x + (b - Ax)$. Apply the Leray–Schauder fixed-point theorem to find a condition ensuring existence of a solution of $Ax = b$. Explain your condition in terms of the usual solvability conditions of linear algebra. Use your answer to understand (the rather odd) Lemma* 7 *in Chapter* 4.

Exercise 79. *Reconsider the nonlinear elliptic BVP:*

$$-k\Delta u + |u|u = f \in L^2 \text{ in } \Omega,\ \ u = 0 \text{ on } \partial\Omega.$$

Prove that a solution exists for any $k > 0$ and $f \in L^2(\Omega)$. State carefully any compactness assumption you need in the proof.

Exercise 80 (Newton's method). *Newton's method for solving $f(x) = b$ is*

Guess x_0.

Compute: $r_n = b - f(x_n)$.

Solve: $f'(x_n)\Delta x = r_n$.

Update: $x_{n+1} = x_n + \Delta x$.

Show that if Newton's method is applied to steady NSE, each step requires the solution of (6.4.2) *(where u^* is u_n).*

Exercise 81. *For $f : \mathbb{R}^n \to \mathbb{R}^n$ consider solving $f(x) = b$. Give a definition of a nonsingular and isolated solution x^*. Prove that nonsingular implies isolated.*

Exercise 82 (one picture of transition to turbulence). *Read and explain the description of transition to turbulence through period doubling in Marsden and Hughes* [54] *and Eckman* [30]. *Relate the first stages to the theory in this chapter.*

Exercise 83 (the d'Alembert paradox). *Read about d'Alembert's paradox (e.g., in* [92]*) and explain the original motivation of Oseen for proposing study of the (now called) Oseen problem.*

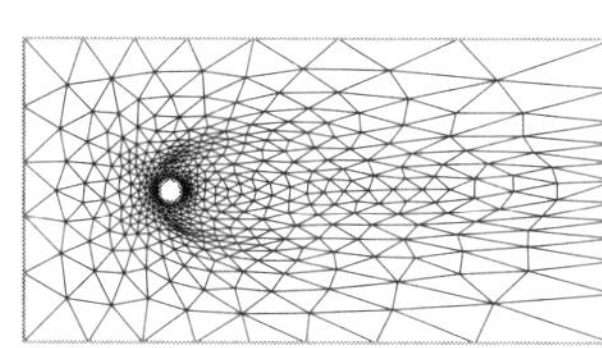

Chapter 7
Approximating Steady Flows

The whole of the development and operations of analysis are now capable of being executed by machines.... As soon as an Analytical Engine exists, it will necessarily guide the future course of science.

Charles Babbage, in *Passages from the Life of a Philosopher*, Longman, Green, Longman, Roberts and Green, London, 1864.

7.1 Formulation and Stability of the Approximation

This chapter considers the problem of approximating a steady solution (u, p) of NSEs:

$$\begin{aligned} &-\nu\Delta u + u \cdot \nabla u + \nabla p = f \text{ in } \Omega, \\ &\nabla \cdot u = 0 \text{ in } \Omega, \\ &u = 0 \text{ on } \partial\Omega \left(\text{and } \int_\Omega p dx = 0 \right). \end{aligned} \tag{7.1.1}$$

The steady NSE present the two interesting problems of stability bounds for the pressure (coupling between $\nabla \cdot u = 0$ and ∇p) and stability bounds for the velocity (due to the nonlinearity). In the methods we consider, boundedness of the pressure by problem data is ensured by using elements which satisfy the discrete inf-sup condition. Remarkably, this condition, necessary and sufficient for stability of the linear Stokes problem, is also sufficient for stability and boundedness of the pressure in the nonlinear NSE. The other main difficulty is the nonlinearity. The solution of this problem which ensures the physical energy bounds hold for approximate solutions is to explicitly skew-symmetrize the nonlinearity in the equations for the approximate solution. We prove the essential stability bounds and by elaborating on this stability proof, show convergence of the approximate solutions. The convergence proof we give is for the case where a globally unique solution exists, i.e., under the small data condition of Chapter 6. This small data condition is thus important to understand. We show in section 4 that the technical condition on smallness of the problem data and associated Reynolds number is the condition on the time-dependent problem which ensures that any perturbation of the steady solution will return exponentially fast to the steady solution.

Recall that the natural variational formulation of (7.1.1) is in (X, Q) or V where

$$X := \{v \in L^2(\Omega)^d : \nabla v \in L^2(\Omega)^{d\times d} \text{ and } v = 0 \text{ on } \partial\Omega\}$$
$$Q := \left\{q \in L^2(\Omega) : \int_\Omega q\, dx = 0\right\},$$

and

$$V := \{v \in X : (q, \nabla \cdot v) = 0\, \forall\, q \in Q\}.$$

It reads: find $(u, p) \in (X, Q)$ satisfying

$$\nu(\nabla u, \nabla v) + (u \cdot \nabla u, v) - (p, \nabla \cdot v) = (f, v)\, \forall\, v \in X, \qquad (7.1.2)$$
$$(\nabla \cdot u, q) = 0, \ \forall\, q \in Q.$$

The solution of the first problem—stability of the pressure—is to use finite element spaces

$$X^h \subset X, \quad Q^h \subset Q$$

satisfying (LBB^h):

$$(LBB^h) \qquad \inf_{q^h\in Q^h} \sup_{v^h\in X^h} \frac{(q^h, \nabla \cdot v^h)}{||q^h||\, ||\nabla v^h||} \geq \beta > 0.$$

The second problem is treated by carefully formulating the nonlinearity in the discrete problem. Recall the vector identity

$$u \cdot \nabla u \cdot u = \frac{1}{2}\nabla \cdot (u|u|^2) - \frac{1}{2}|u|^2 \nabla \cdot u,$$

from which it is clear that $(u \cdot \nabla u, u) = 0$ precisely because $u \in V$, i.e., because $u \cdot \widehat{n}$ vanishes *exactly* on the boundary and $\nabla \cdot u$ is *exactly* zero. Difficulties can arise because an approximate solution

$$u^h \in V^h := \{v \in X^h : (q^h, \nabla \cdot v^h) = 0\, \forall\, q^h \in Q^h\}$$

is approximately but (except in very special cases) never exactly divergence free:

$$\nabla \cdot u^h \neq 0 \text{ and } V^h \not\subset V \quad \text{in general!}$$

To formulate the discrete problem so as to eliminate any such potential difficulties, define the (explicitly skew symmetrized) trilinear form

$$b^*(u, v, w) := \frac{1}{2}\,(u \cdot \nabla v, w) - \frac{1}{2}(u \cdot \nabla w, v).$$

Recall from Chapter 6 that the nonlinearity is continuous on $X \times X \times X$ (and thus on $V \times V \times V$ as well). We have associated with the nonlinearity the finite, continuity constants (which depend on the domain Ω):

$$M = M(\Omega) := \sup_{u,v,w\in X} \frac{(u \cdot \nabla v, w)}{||\nabla u||\, ||\nabla v||\, ||\nabla w||} < \infty,$$

$$N = N(\Omega) := \sup_{u,v,w\in V} \frac{(u \cdot \nabla v, w)}{||\nabla u|| ||\nabla v|| ||\nabla w||} < \infty.$$

Since $V \subset X$ it follows that $N \le M(< \infty)$.

Lemma 18 (skew-symmetry and continuity). *For $u \in V$, $v, w \in X$,*

$$(u \cdot \nabla v, w) = b^*(u, v, w).$$

Further,

$$b^*(u, v, v) = 0 \text{ for any } u, v \in X, \text{ and}$$

$$b^*(u, v, w) \le M\, ||\nabla u||\, ||\nabla v||\, ||\nabla w|| \;\forall u, v, w \in X,$$

for the same $M = M(\Omega)$.

Proof. Exercise. □

Thus, another variational formulation of a solution of (7.1.2) which is equivalent for the continuous problem is as follows: find $(u, p) \in (X, Q)$ satisfying

$$\begin{aligned} &(\nu \nabla u, \nabla v) + b^*(u, u, v) - (p, \nabla \cdot v) = (f, v)\; \forall\, v \in X, \\ &(q, \nabla \cdot u) = 0\; \forall\, q \in Q. \end{aligned} \tag{7.1.3}$$

The finite element approximations $(u^h, q^h) \in (X^h, Q^h)$ that we consider satisfy

$$\begin{aligned} &(\nu \nabla u^h, \nabla v^h) + b^*(u^h, u^h, v^h) - (p^h, \nabla \cdot v^h) = (f, v^h)\; \forall\, v^h \in X^h, \\ &(q^h, \nabla \cdot u^h) = 0\; \forall\, q^h \in Q^h. \end{aligned} \tag{7.1.4}$$

The formulation (7.1.4) can be written as follows: find $u^h \in V^h$ satisfying

$$(\nu \nabla u^h, \nabla v^h) + b^*(u^h, u^h, v^h) = (f, v^h)\; \forall\, v^h \in V^h. \tag{7.1.5}$$

Lemma 19 (stability). *The finite element approximation* (7.1.4) *is stable,*

$$\nu ||\nabla u^h|| \le ||f||_{-1},$$

and if (LBB^h) holds, then

$$||p^h|| \le \beta^{-1}(2 + M\nu^{-2}||f||_{-1})||f||_{-1}.$$

Proof. For the first result set $v^h = u^h$ in (7.1.5). This is equivalent to setting $v^h = u^h$ and $q^h = p^h$ in (7.1.4) and adding. Using $b^*(u^h, u^h, u^h) = 0$ we get

$$\nu ||\nabla u^h||^2 = (f, u^h) \le ||f||_{-1}\, ||\nabla u^h||,$$

giving the first inequality. For the stability bound of the pressure, solve (7.1.4) for $(p^h, \nabla \cdot v^h)$:

$$\begin{aligned} (p^h, \nabla \cdot v^h) &= (f, v^h) - \nu(\nabla u^h \cdot \nabla v^h) - b^*(u^h, u^h, v^h), \text{ so} \\ (p^h, \nabla \cdot v^h) &\le ||f||_*\, ||\nabla v^h|| + \nu ||\nabla u^h||\, ||\nabla v^h|| + M\, ||\nabla u^h||\, ||\nabla u^h||\, ||\nabla v^h||. \end{aligned}$$

Thus,

$$\frac{(p^h, \nabla \cdot v^h)}{||\nabla v^h||} \leq ||f||_* + \nu||\nabla u^h|| + M\,||\nabla u^h||^2.$$

Taking the supremum over $v^h \in X^h$ and using $\nu||\nabla u^h|| \leq ||f||_{-1}$ and the assumed (LBB^h) condition gives the pressure bound. □

Our error analysis in section 7.3 will be based on small data assumption that the solution of (7.1.1) is globally stable, in the form

$$M\,\nu^{-2}\,||f||_{-1} \leq \alpha < 1. \tag{7.1.6}$$

As we describe below, it is not technically hard (although it takes longer) to sharpen the error analysis to replace the above small data condition (7.1.6) by the small data condition $N\nu^{-2}\,||f||_* \leq \alpha < 1$ from Chapter 6.[58]

It is useful to think of the error analysis we develop in this chapter as a combination of three ideas:

- The proof of Céa's lemma shows how to handle the contribution of the viscous terms to the error.
- The (LBB^h) condition shows how to handle the coupling between $\nabla \cdot u = 0$ and ∇p in the error analysis.
- The small data condition (7.1.6) will be used to control the nonlinearity in the error analysis.

Remark 7. *Define the finite constant*

$$N^h = N^h(\Omega) := \sup_{u^h, v^h, w^h \in V^h} \frac{(u^h \cdot \nabla v^h, w^h)}{||\nabla u^h||\,||\nabla v^h||\,||\nabla w^h||} < \infty.$$

The error analysis is easily generalized to the case when M is replaced by the above sharper constant N^h in the small data assumption. It can also be shown that

$$N^h \to N, \textit{as } h \to 0.$$

(This is an interesting demonstration. It is proved in [44].*) Thus, M can also be replaced by N in the small data assumption* (7.1.6) *by adding the extra hypothesis that h is sufficiently small.*

7.2 A Simple Example

> *He (Gill) climbed up and down the lower half of the rock over and over, memorizing the moves.... He says that "going up and down, up and down eventually... your mind goes blank and you climb by well cultivated instinct."*
>
> J. Krakauer, from his book *Eiger Dreams*.

[58] For this sharpening, the constant M (which is defined with supremum taken over X) must be replaced with N (defined with supremum over V) and the norm on X^* by the norm on V^* (i.e., $||f||_{-1}$ replaced by $||f||_*$).

The error analysis we are embarked upon in this chapter is complex but the underlying ideas are simple. Recall from Chapter 6 that global uniqueness under a small data condition often has a contractive map implicit somewhere in the formulation. Thus, to give you some feel for the analysis we first wish to give a simple example isolating the way a contractive nonlinearity is handled.

Consider approximating *fixed points of a contraction.* Suppose

$$T : X \to X \text{ satisfies } ||T(x) - T(y)||_X \leq \alpha ||x - y||_X \ \forall\, x, y \in X,$$

for some $\alpha < 1$. Let $X^h \subset X$ be a subspace. Let x^* be the unique fixed point of T:

$$x^* - T(x^*) = 0.$$

Let $x^h \in X^h$ be the Galerkin approximation to x^*

$$(x^h - T(x^h), y^h)_X = 0 \ \forall y^h \in X^h.$$

Let's consider the error in x^* now. Clearly,

$$((x^* - x^h) - (T(x^*) - T(x^h)), y^h)_X = 0.$$

Thus,

$$((x^* - x^h), y^h)_X = (T(x^*) - T(x^h), y^h)_X.$$

Let $P^h : X \to X^h$ denote the orthogonal projection into X^h. Set $y^h = P^h(x^* - x^h)$. Then,

$$||P^h(x^* - x^h)||_X^2 \leq \alpha ||x^* - x^h||_X ||P^h(x^* - x^h)||_X$$

Thus, $||P_{X^h}(x^* - x^h)||_{X^h} \leq \alpha ||x^* - x^h||_X$. The Pythagorean theorem then implies

$$\begin{aligned} ||x^* - x^h||_X^2 &= ||x^* - P^h x^*||_X^2 + ||P^h(x^* - x^h)||_X^2 \\ &\leq ||x^* - P^h x^*||_X^2 + \alpha^2 ||x^* - x^h||_X^2. \end{aligned}$$

Collecting terms proves the following.

Proposition 12 (error in Galerkin approximation of fixed points). *Let x^* be the unique fixed point of a contractive map in X with contraction constant α and let x^h denote its Galerkin approximation. The error satisfies*

$$||x^* - x^h||_X \leq (1 - \alpha^2)^{-\frac{1}{2}} \inf_{w^h \in X^h} ||x^* - w^h||_X.$$

7.3 Errors in Approximations of Steady Flows

We now turn to the analysis of the error in the NSE under the global uniqueness condition

$$\nu^{-2} M \, ||f||_{-1} \leq \alpha < 1. \tag{7.3.1}$$

Recall that this condition ensures that the continuous problem can be formulated as a fixed point of a contractive map in V, the space of divergence-free functions.[59] We shall first see that exactly this condition also ensures u^h to be a fixed point of a contractive map in

$$V^h := \{v^h \in X^h : (q^h, \nabla \cdot v^h) = 0 \; \forall \, q^h \in Q^h\},$$

which is the space of *discretely* divergence free functions.

We can avoid a common mistake at this point by remembering that

$$V^h \not\subset V \text{ (in general)},$$

so the error analysis *must* contain extra terms to account for discretely divergence free functions not being exactly divergence-free.

Consider the formulation in $V^h : u^h \in V^h$ satisfies

$$\nu(\nabla u^h, \nabla v^h) + b^*(u^h, u^h, v^h) = (f, v) \; \forall \, v^h \in V^h. \tag{7.3.2}$$

Lemma 20 (stability of the velocity). *Any solution u^h of the above satisfies*

$$\nu||\nabla u^h|| \leq ||f||_{-1}.$$

Proof. For the proof, set $v^h = u^h$ and use the fact that $b^*(\cdot, \cdot, \cdot)$ has been explicitly skew-symmetrized. $\square$

Using this á priori bound it is possible to follow almost exactly the existence proof of the continuous case.

Proposition 13 (existence of the discrete velocity and pressure). *For any $f \in L^2(\Omega)$, there exists at least one solution u^h. If (X^h, Q^h) satisfies (LBB^h), p^h also exists.*

Uniqueness can also be proven by an analogous argument. Define the finite constant N^h and $(V^h)^*$ norm by

$$N^h := \sup_{u^h, v^h, w^h \in V^h} \frac{b^*(u^h, v^h, w^h)}{||\nabla u^h|| \, ||\nabla v^h|| \, ||\nabla w^h||},$$

$$||f||_{h,*} := \sup_{v^h \in V^h} \frac{(f, v^h)}{||\nabla v^h||}.$$

It is easy to show that $N^h \leq M < \infty$ and $||f||_{h,*} \leq ||f||_{-1}$.

Proposition 14 (uniqueness under a small data condition). *Suppose the small data condition*

$$\nu^{-2} N^h ||f||_{h,*} \leq \alpha < 1$$

[59] This condition is the stronger of the two given in Chapter 6 for global uniqueness. The convergence theory in this section can be extended to the weaker condition by adding the assumption that the mesh width h is small enough. We work with this one because it allows a shorter and more direct approach in the error analysis which makes the key ideas and critical steps clear.

holds. Then, there is at most one solution of (7.3.2). *Uniqueness also holds with N^h replaced by M and $||f||_{h,*}$ by $||f||_{-1}$.*

Proof. Let u_1^h, u_2^h be two solutions and let $w^h = u_1^h - u_2^h$. By subtraction, $w^h \in V^h$ satisfies

$$\nu(\nabla w^h, \nabla v^h) + b^*(u_1^h, u_1^h, v^h) - b^*(u_2^h, u_2^h, v^h) = 0 \; \forall \, v^h \in V^h.$$

Setting $v^h = w^h$ and adding and subtracting terms gives

$$\nu||\nabla w^h||^2 + b^*(u_1^h, w^h, w^h) + b^*(w^h, u_2^h, w^h) = 0.$$

The second term vanishes because $b^*(\cdot, \cdot, \cdot)$ is explicitly skew-symmetrized. Thus,

$$\begin{aligned}\nu||\nabla w^h||^2 = -b^*(w^h, u_2^h, w^h) &\le N^h ||\nabla u_2^h|| \; ||\nabla w^h||^2 \\ &\le N^h \, \nu^{-1} ||f||_{h,*} \, ||\nabla w^h||^2.\end{aligned}$$

Thus, collecting terms and dividing through by ν,

$$(1 - N^h \nu^{-2} ||f||_{h,*}) \, ||\nabla w^h||^2 \le 0.$$

Uniqueness follows since the small data condition implies that the multiplier on the left-hand side is positive. □

Remark 8. *It is easy to show $N^h \le M$ so the continuous uniqueness condition* (7.3.1) *implies the discrete one. Further, if X^h becomes dense in X as $h \to 0$ and (LBB^h) holds then $N^h \to N$ as $h \to 0$* [44].

Next, we will need to derive the error equation. This is the nonlinear equivalent of Galerkin orthogonality. Since $V^h \not\subset V$, this means subtracting (7.3.2) from (7.1.2). This gives $\forall \, v^h \in V^h$ and $\forall \, q^h \in Q^h$,

$$\begin{aligned}(\nu\nabla(u - u^h), \nabla v^h) + b^*(u, u, v^h) - b^*(u^h, u^h, v^h) + (p, \nabla \cdot v^h) = 0, \\ \nabla \cdot u = 0 \text{ and } (\nabla \cdot u^h, q^h) = 0.\end{aligned}$$

It's an important refinement of this equation to note that since $v^h \in V^h$, $(\nabla \cdot v^h, q^h) = 0$ and we can write

$$(p, \nabla \cdot v^h) = (p - q^h, \nabla \cdot v^h) \; \forall \, q^h \in Q^h \text{ and } v^h \in V^h.$$

The error equation then becomes

$$\begin{aligned}(\nu\nabla(u - u^h), \nabla v^h) + b^*(u, u, v^h) - b^*(u^h, u^h, v^h) \\ + (p - q^h, \nabla \cdot v^h) = 0 \; \forall \, v^h \in V^h, q^h \in Q^h.\end{aligned} \tag{7.3.3}$$

This is a complicated, nonlinear error equation. Yet, we can almost get to the final estimate by following our noses:

1. Write $u - u^h = (u - v^h) - (u^h - v^h) = \eta - \phi^h$ (where $\eta = u - v^h$ and $\phi^h = (u^h - v^h) \in V^h$), where v^h is an optimal approximation of u in V^h (see Lemma 9 of Chapter 4).

2. Put the ϕ^h terms on one side and the η terms on the other; set $v^h = \phi^h$ and try to get a bound of $||\nabla\phi^h||$ in terms of $||\nabla\eta||$.

3. Quadratic nonlinearities are always treated the same way:[60]

$$\begin{aligned} aa - bb &= aa - ab + ab - bb \\ &= a(a-b) + (a-b)b. \end{aligned}$$

4. Applying Cauchy–Schwarz inequalities to the right-hand side we will get "good" (η) terms and "bad" (ϕ^h) terms. Use the small data condition to hide (dominate/subsume) these bad terms into the good terms of the left-hand side.

5. Apply the triangle inequality

$$||\nabla(u-u^h)|| \le ||\nabla\eta|| + ||\nabla\phi^h||.$$

6. Take the infimum over $v^h \in V^h$.

Let us apply these six steps and look for subtleties along the way. Step 1 gives $(u - u^h = \eta - \phi^h)$,

$$\begin{aligned} \nu(\nabla\phi^h, \nabla v^h) = &\operatorname{Re}^{-1}(\nabla\eta, \nabla v^h) \\ &+ b^*(u,u,v^h) - b^*(u^h,u^h,v^h) + (p - q^h, \nabla\cdot v^h) \end{aligned}$$

for any $v^h \in V^h$ and $q^h \in Q^h$. Setting $v^h = \phi^h$ gives

$$\nu||\nabla\phi^h||^2 = \operatorname{Re}^{-1}(\nabla\eta, \nabla\phi^h) + b^*(u,u,\phi^h) - b^*(u^h,u^h,\phi^h) + (p-q^h, \nabla\cdot\phi^h). \quad (7.3.4)$$

Consider the nonlinear terms. Using the idea of step 3 gives

$$\begin{aligned} &b^*(u,u,\phi^h) - b^*(u^h,u^h,\phi^h) \\ &= b^*(u, u-u^h, \phi^h) + b^*(u-u^h, u^h\phi^h) \\ &= b^*(u, \eta-\phi^h, \phi^h) + b^*(\eta-\phi^h, u^h, \phi^h) \\ &(\text{since } b^*(u,\phi^h,\phi^h) = 0 \text{ by skew-symmetry of } b^*(\cdot,\cdot,\cdot)) \\ &= b^*(u,\eta,\phi^h) + b^*(\eta,u^h,\phi^h) - b^*(\phi^h,u^h,\phi^h). \end{aligned}$$

Thus,

$$\begin{aligned} &|b^*(u,u,\phi^h) - b^*(u^h,u^h,\phi^h)| \\ &\le M||\nabla u||\,||\nabla\eta||\,||\nabla\phi^h|| + M||\nabla u^h||\,||\nabla\eta||\,||\nabla\phi^h|| + M||\nabla u^h||\,||\nabla\phi^h||^2. \end{aligned}$$

Inserting this bound into (7.3.4) gives

$$\begin{aligned} &(\nu - M||\nabla u^h||)\,||\nabla\phi^h||^2 \\ &\le \nu(\nabla\eta, \nabla\phi^h) + (p-q^h, \nabla\cdot\phi^h) \\ &+ M(||\nabla u|| + ||\nabla u^h||)\,||\nabla\eta||\,||\nabla\phi^h||. \end{aligned}$$

[60] Actually, in the vector case $ab \neq ba$, so there are two possibilities: using ab and ba in this splitting.

On the right-hand side, first the pressure term is bounded by $(p-q^h, \nabla\cdot\phi^h) \le ||p-q^h||||\nabla\cdot\phi^h||$ and $||\nabla\cdot\phi^h|| \le \sqrt{d}||\nabla\phi^h||$. Next, we now use the idea of step 4 ($ab \le \epsilon a^2 + \frac{1}{4\epsilon}b^2$):

$$\begin{aligned}
&(\nu - M||\nabla u^h||)||\nabla\phi^h||^2\\
&\le \epsilon||\nabla\phi^h||^2 + \frac{\nu^{-2}}{4\epsilon}||\nabla\eta||^2\\
&+\epsilon||\nabla\phi^h||^2 + \frac{d}{4\epsilon}||p-q^h||^2\\
&+\epsilon||\nabla\phi^h||^2 + \frac{M^2}{4\epsilon}(||\nabla u|| + ||\nabla u^h||)^2||\nabla\eta||^2.
\end{aligned}$$

Collecting terms gives

$$\begin{aligned}
&(\nu - M||\nabla u^h|| - 3\epsilon)||\nabla\phi^h||^2\\
&\le \nu^2/4\epsilon||\nabla\eta||^2\\
&+\frac{d}{4\epsilon}||p-q^h||^2 + \frac{M^2}{4\epsilon}(||\nabla u|| + ||\nabla u^h||)^2||\nabla\eta||^2.
\end{aligned}$$

Lemma 20 gives $\nu||\nabla u^h|| \le ||f||_{-1}$ and the small data condition (7.3.1) implies

$$M||\nabla u^h|| \le M\text{Re}||f||_{-1} \le \alpha\text{Re}^{-1} < \text{Re}^{-1}.$$

Thus, we have, picking $\epsilon = \frac{1-\alpha}{6}\,\nu\,||\nabla\phi^h||$,

$$\begin{aligned}
&\frac{\nu(1-\alpha)}{2}||\nabla\phi^h||^2 = [\nu - \alpha\nu - 3\epsilon]||\nabla\phi^h||^2\\
&\le \frac{3\nu}{2(1-\alpha)}||\nabla\eta||^2 + \frac{3d}{2(1-\alpha)}\nu^{-1}||p-q^h||^2\\
&+\frac{3}{2(1-\alpha)}\nu^{-1}(2\nu^{-1}||f||_*)^2||\nabla\eta||^2.
\end{aligned}$$

Collecting terms gives

$$\begin{aligned}
&||\nabla\phi^h||^2 \le \frac{3}{(1-\alpha)^2}||\nabla\eta||^2 + \frac{3d}{(1-\alpha)^2}\,\nu^{-2}||p-q^h||^2\\
&+\frac{12}{(1-\alpha)^2}\,\nu^{-4}||f||_{-1}^2||\nabla\eta||^2.
\end{aligned}$$

The triangle inequality now yields the error estimate in the following theorem.

Theorem 20 (convergence of the FEM). *Suppose the global uniqueness condition* (7.3.1) *holds. Then*

$$||\nabla(u-u^h)|| \le C(\nu, f)\left\{\inf_{v\in V^h}||\nabla(u-v^h)|| + \inf_{q\in Q^h}||p-q^h||\right\}.$$

If condition (LBB^h) *holds, the infimum over V^h can be replaced by one over X^h, giving*

$$||\nabla(u-u^h)|| \le C(\nu, f, \beta^h)\left\{\inf_{v\in X^h}||\nabla(u-v^h)|| + \inf_{q\in Q^h}||p-q^h||\right\}.$$

Proof. The first inequality is proved already. For the second, pick (v^h, q^h) to be the Stokes projection of (u, q) in (X^h, Q^h) and use Lemma 9 of Chapter 4. □

For specific choices of div-stable elements, rates of convergence can be determined.

Corollary 4 (rates of convergence for the MINI element). *Suppose that the global uniqueness condition* (7.3.1) *holds and let* X^h, Q^h *be the MINI element. Then,*

$$||\nabla(u - u^h)|| \leq C(f, \nu)h\{||\nabla\nabla u|| + ||\nabla p||\}.$$

If additionally Ω *is a convex, planar polygon, and* $f \in L^2(\Omega)$, *then*

$$||\nabla(u - u^h)|| \leq C(f, \nu)h.$$

Proof. The first inequality follows by using interpolation theory for linear elements from Chapter 2. The second inequality needs one additional ingredient: regularity theory for the Stokes (and Navier–Stokes) equation. In the indicated setting, it is known that solutions of the Stokes problem satisfy

$$||\nabla\nabla u|| + ||\nabla p|| \leq C(\Omega)||f||.$$

It is possible to prove the same about the steady NSE using a bootstrap argument as follows. Rewrite the NSE as a Stokes problem with the right-hand side $f - u \cdot \nabla u$. The above regularity result implies

$$||\nabla\nabla u|| + ||\nabla p|| \leq C(\Omega, \nu)||f - u \cdot \nabla u|| \leq C(||f|| + ||u \cdot \nabla u||).$$

Now, by the Cauchy–Schwarz and Ladyzhenskaya inequalities,

$$\begin{aligned} ||u \cdot \nabla u||^2 &= \int_\Omega |u \cdot \nabla u|^2 dx \leq \int_\Omega |u|^2 |\nabla u|^2 dx \\ &\leq \left(\int_\Omega |u|^4 dx\right)^{\frac{1}{2}} \left(\int_\Omega |\nabla u|^4 dx\right)^{\frac{1}{2}} = ||u||^2_{L^4(\Omega)} ||\nabla u||^2_{L^4(\Omega)}. \end{aligned}$$

Thus, by Theorem 4 of Chapter 1,

$$||u \cdot \nabla u|| \leq C||u|| ||\nabla u||^{\frac{1}{2}} ||\nabla\nabla u||^{\frac{1}{2}},$$

and using Exercise 1 of Chapter 1 with $p = 4$ and $q = 4/3$ gives, for any $\varepsilon > 0$,

$$||u \cdot \nabla u|| \leq \frac{\varepsilon}{4}||\nabla\nabla u||^2 + C(\varepsilon)(||u|| ||\nabla u||^{\frac{1}{2}})^{\frac{4}{3}} \leq \frac{\varepsilon}{4}||\nabla\nabla u||^2 + C(\varepsilon, \Omega)||\nabla u||^2.$$

Thus, we have for the NSE solution

$$\begin{aligned} ||\nabla\nabla u|| + ||\nabla p|| &\leq C(\Omega, \nu)||f - u \cdot \nabla u|| \leq C(||f|| + ||u \cdot \nabla u||) \\ &\leq C||f|| + \frac{\varepsilon}{4}||\nabla\nabla u||^2 + C(\varepsilon, \Omega)||\nabla u||^2, \end{aligned}$$

and it follows that

$$||\nabla\nabla u|| + ||\nabla p|| \leq C(\Omega, \nu)||f||,$$

which completes the proof. □

Corollary 5 (rates of convergence for the Hood–Taylor element). *Suppose that the global uniqueness condition* (7.3.1) *holds and let $X^h_?$ Q^h be the Hood–Taylor element. Then,*

$$||\nabla(u-u^h)|| \leq C(f,\nu)h^2\{||\nabla\nabla\nabla u|| + ||\nabla\nabla p||\}.$$

Proof. This follows like the previous corollary. □

Under the discrete inf-sup condition optimal error estimates for the pressure also hold (Exercise 94). The proof is almost identical to the Stokes problem.

7.4 More on the Global Uniqueness Conditions

> *For nonlinear equations, such as the Navier–Stokes equations, it is known that a regular solution for the nonstationary problem need not exist for all times $t \geq 0$, it need not converge towards the solution of the stationary problem as $t \to \infty$, when the boundary conditions and the forces converge towards a stationary solution.*
>
> O. A. Ladyzhenskaya, in [62].

> *There is scarcely any question in dynamics more important for Natural Philosophy than the stability of motion.*
>
> W. Thompson and P. G. Tait, in *Treatise on Natural Philosophy*, Oxford, 1867, p. 346.

The error analysis in this chapter was based on a global uniqueness assumption on the data and (through that) the Reynolds number. Since fluid flow is inherently time-dependent, it is important to understand what this condition means for the time-dependent NSE. To this end, let $(u^*(x), p^*(x))$ be a solution of the (nondimensionalized) steady NSE:

$$\begin{aligned}&-\mathrm{Re}^{-1}\Delta u^* + u^*\cdot\nabla u^* + \nabla p^* = f(x) \text{ in } \Omega,\\ &\nabla\cdot u^* = 0 \text{ in } \Omega,\\ &u^* = 0 \text{ on } \partial\Omega \text{ and } (p,1) = 0.\end{aligned} \tag{7.4.1}$$

Under the small data condition for global uniqueness,

$$\mathrm{Re}^2\, N||f||_* < 1, \tag{7.4.2}$$

the solution of (7.4.1) is unique and finite element approximations converge to (u^*, p^*) as $h \to 0$. It is worthwhile to try to understand the meaning of the above global uniqueness condition in the time-dependent problem. Accordingly, let $(u(x,t), p(x,t))$ be the solution of the time-dependent problem with the same $f(x)$:

$$\begin{aligned}&u_t - \mathrm{Re}^{-1}\Delta u + u\cdot\nabla u + \nabla p = f(x) \text{ in } \Omega,\\ &\nabla\cdot u = 0 \text{ and } u(x,0) = u_0(x) \text{ in } \Omega\\ &u = 0 \text{ on } \partial\Omega \text{ and } (p,1) = 0.\end{aligned} \tag{7.4.3}$$

Let $w(x,t) = u(x,t) - u^*(x)$. Subtracting the variational formulations of (7.3.2) and (7.4.3) gives that $w : [0,\infty) \to V$ satisfies

$$(w_t, v) + (u\cdot\nabla u, v) - (u^*\cdot\nabla u^*, v) + \mathrm{Re}^{-1}(\nabla w, \nabla v) = 0\; \forall\, v \in V,$$

$$w(x,0) = u(x,0) - u^*(x).$$

Adding and subtracting terms and setting $v = w$ gives

$$\begin{aligned}
&\frac{1}{2}\frac{d}{dt}||w||^2 + \mathrm{Re}^{-1}||\nabla w||^2 = -(u \cdot \nabla w, w) - (w \cdot \nabla u^*, w) \\
&= 0 - (w \cdot \nabla u^*, w) \\
&\leq N||\nabla w||\;||\nabla u^*||\;||\nabla w|| \\
&\leq \text{using the bound on } u^* \leq N\ \mathrm{Re}||f||_*||\nabla w||^2.
\end{aligned}$$

Thus,

$$\frac{1}{2}\frac{d}{dt}||w||^2 + \mathrm{Re}^{-1}(1 - N\ \mathrm{Re}^2||f||_*)||\nabla w||^2 \leq 0.$$

Now the small data condition implies

$$N\ \mathrm{Re}^2||f||_* \leq \alpha < 1 \ or\ 1 - N\ \mathrm{Re}^2||f||_* \geq 1 - \alpha > 0.$$

Thus, by the Poincaré–Friedrichs inequality

$$\frac{d}{dt}\ ||w||^2 \leq -2\ \mathrm{Re}^{-1}(1-\alpha)||\nabla w||^2 \leq -\delta||w||^2$$

where $\delta = 2\ \mathrm{Re}^{-1}(1-\alpha)\ C_{PF} > 0$. Integration gives

$$||w||^2(t) \leq e^{-\delta t}||w(0)|| \to 0 \text{ as } t \to \infty$$

and

$$||u(x,t) - u^*(x)|| \to 0$$

exponentially fast as $t \to \infty$.

Thus, *the global uniqueness condition for the steady problem is really a condition for global, exponential, asymptotic stability of the (unique) steady solution in the time-dependent problem.*

7.5 Remarks on Chapter 7

Our presentation of the error analysis in the uniqueness case is a simplification of that in the fundamental book of Girault and Raviart [44]. The last section, treating physical stability of the steady solution of the time-dependent problem, hints at the vast literature on stability of fluid motion. See Galdi [41] and Straughan [94] and the references therein for more of this beautiful area. Fluid flow typically is three-dimensional and time-dependent. When working on simplified models (e.g., steady NSE, boundary layer equations) or settings (e.g., two dimensions), the meaning of every condition imposed needs to be understood in the context of the three-dimensional, time-dependent Navier–Stokes equations.

When the Reynolds number is larger, the global uniqueness condition no longer holds. It is known, however, that solutions are generically nonsingular. A convergence theory for finite element methods is also in place for this case as well. The error analysis in the nonuniqueness case is the fundamental next step beyond the presentation of this chapter. It is explained lucidly in [46], and the full technical details of two complementary approaches are given in [44] and [45].

A posteriori error analysis of nonlinear problems can be performed in the small (that is, beginning with the assumption that the approximate solution is close enough to the true solution), e.g., Verfürth [100]. This assumption is reasonable once á priori error estimates are known. Ideally, complete knowledge precedes computations; practically, for nonlinear problems it does not. There are many problems for which insight into the behavior of the numerical method can be obtained from one type of estimate, and other types of estimates are unavailable until numerical calculations give more insight into the fluid velocity. For nonlinear problems, analytical insight and computational experience must grow together or both will stagnate.

7.6 Exercises

Exercise 84 (another useful skew-symmetrization). *Define* $b^{**}(u, v, w) := (u \cdot \nabla v, w) + \frac{1}{2}((\nabla \cdot u)v, w)$. *Show that* $b^{**}(\cdot, \cdot, \cdot)$ *is skew symmetric on* X *and prove continuity of* $b^{**}(u, v, w)$.

Exercise 85. *Suppose* (LBB^h) *holds. Define*

$$N^h := \sup_{u^h, v^h, w^h \in V^h} \frac{(u^h \cdot \nabla v^h, w^h)}{||\nabla u^h||\ ||\nabla v^h||\ ||\nabla w^h||}.$$

Look up and present the proof in [44] *that* $N^h \to N$ *as* $h \to 0$, *provided* (X^h, Q^h) *becomes dense in* (X, Q) *as* $h \to 0$.

Exercise 86. *Let* x^* *be the fixed point of a contractive map:* $x^* = T(x^*)$ *in* X. *Show that its Galerkin approximation in* $X^h \subset X$ *exists and is unique.*

Exercise 87 (a reaction-diffusion problem). (a) *Let* $f(x) \in L^2(\Omega)$, $\epsilon > 0$, *and consider the nonlinear BVP: find* $u(x)$ *satisfying*

$$-\Delta u = f(x) + \epsilon\ g(u) \text{ in } \Omega,\ \ u = 0 \text{ on } \partial\Omega. \tag{7.6.1}$$

Let $g : \mathbb{R} \to \mathbb{R}$ *be globally Lipschitz:*

$$|g(u) - g(v)| \le L|u - v|\ \forall\, u, v \in \mathbb{R}.$$

Let $X := \{u \in L^2(\Omega) : \nabla u \in L^2(\Omega) \text{ and } u = 0 \text{ on } \partial\Omega\}$. *Find the variational formulation of* (7.6.1) *in* X.

(b) *Let* $T(f) = w$ *be the solution of:* $w \in X$ *satisfies*

$$(\nabla w, \nabla v) = (f, v)\ \forall\ v \in X.$$

Show (7.6.1) *is equivalent to the fixed point problem: find* $u \in X$ *with*

$$u = \mathcal{F}(u) := T(f + \epsilon\ g(u)).$$

Reconsider (7.6.1). *Show* $\mathcal{F}(\cdot)$ *is a contraction for* ϵ *small enough using the equivalent inner product and norm*

$$(u, v)_X := (\nabla u, \nabla v)_{L^2(\Omega)} \text{ and } ||u||_X = ||\nabla u||.$$

(c) *Let* $P^h : X \to X^h$ *be the orthogonal projection with respect to*

$$(\cdot, \cdot)_X := (\nabla u, \nabla v)_{L^2(\Omega)}.$$

Show that the Galerkin approximation $u^h \in X^h$ *to* (7.6.1) *satisfies*

$$u^h = P^h \mathcal{F}(u^h).$$

Prove an error estimate for $u - u^h$ *for small* ϵ.

(d) *Reconsider*

$$-\Delta u = f(x) + \epsilon\, g(u) \text{ in } \Omega, \; u = 0 \text{ on } \partial\Omega,$$

for ϵ *small. Prove an error estimate for* $u - u^h$ *by direct assault. Mimic one of the proofs of Céa's lemma and use the Lipschitz condition and* ϵ *small to hide the extra terms in the right-hand side arising from the nonlinearity.*

Exercise 88. *Prove an error bound on* $||p - p^h||$ *assuming* (LBB^h) *holds.*

Exercise 89 (lagging the nonlinearity). *Consider the problem of solving the nonlinear system for* (u^h, p^h). *Show that the fixed point iteration*

$$\nu(\nabla u^h_{n+1}, \nabla v^h) + b^*(u^h_n, u^h_n, v^h) = (f, v^h) \;\forall\, v^h \in V^h$$

converges locally under a (very) small data condition and that each step requires the solution of a discrete Stokes problem.

Exercise 90 (a (better) fixed point iteration). *Analyze convergence of the iteration*

$$\nu(\nabla u^h_{n+1}, \nabla v^h) + b^*(u^h_n, u^h_{n+1}, v^h) = (f, v^h) \;\forall\, v^h \in V^h.$$

Find the small data condition sufficient for convergence and compare it with the condition arising in the previous exercise.

Exercise 91 (another skew-symmetrization). *Let* (u^h, p^h) *be the finite element approximation to the steady NSEs using* $b^{**}(\cdot,\cdot,\cdot)$ *(defined in Exercise* 84*) instead of* $b^*(\cdot,\cdot,\cdot)$. *Prove a stability bound for the method. Under a small data condition, prove convergence.*

Exercise 92. *Write down Newton's method in* V^h *for calculating* $u^h \in V^h$. *From this, find its equivalent formulation in* (X^h, Q^h).

Exercise 93. *Consider calculating* u^* *by solving* $u(x, t)$ *and time-stepping to* $t = \infty$ *via the following: given* $\Delta t > 0$ *and* u_0 *calculate* $(u^{n+1} \in V)$ *via*

$$\left(\frac{u^{n+1} - u^n}{\Delta t}, v\right) + \mathrm{Re}^{-1}(\nabla u^{n+1}, \nabla v) + b^*(u^n, u^{n+1}, v) = (f, v) \;\forall v \in V.$$

Prove that u^{n+1} *exists for any* n. *Show* $||u^n - u^*|| \to 0$ *as* $n \to \infty$ *under the small data condition. Is a condition on* Δt *needed?*

Exercise 94 (pressure errors). *Prove an error estimate for the error in the pressure under the discrete inf-sup condition.*

Part III

Time-Dependent Fluid Flow Phenomena

Chapter 8

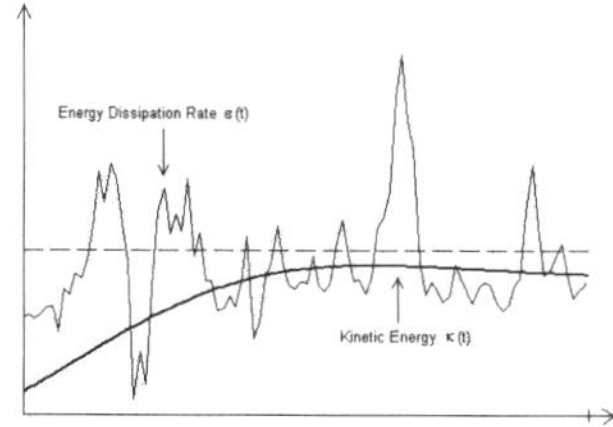

The Time-Dependent Navier–Stokes Equations

As far as I can see, there is today no reason not to regard the hydrodynamic equations as the exact expression of the laws that rule the motion of real fluids.

H. Helmholtz, 1873, in *Über ein Theorem, geometrische ähnliche Bewegungen flüssigkeiten*, Königliche Akademie der Wissenschaffen zu Berlin, Monatsberichte, pp. 501–514.

8.1 Introduction

Consider the flow of a fluid in a region Ω in $\mathbb{R}^2$ or $\mathbb{R}^3$, bounded by walls and driven by a body force $f(x,t)$. The fluid velocity and pressure are functions $u(x,t)$ for $x \in \bar{\Omega}$, $0 \le t \le T$, and $p(x,t)$, for $x \in \Omega$ and $0 < t \le T$, which satisfy

$$\begin{aligned} & u_t + u \cdot \nabla u - \nu \Delta u + \nabla p = f, \ x \in \Omega, \ 0 < t \le T, \qquad (8.1.1) \\ & \nabla \cdot u = 0, \ x \in \Omega \text{ for } 0 < t \le T, \ u(x,0) = u_0(x), \ x \in \Omega, \\ & u = 0 \text{ on } \partial\Omega, \ \int_\Omega p\, dx = 0 \text{ for } 0 < t \le T. \end{aligned}$$

The key idea in making progress in the mathematical understanding of the NSE is the notion of weak solution, introduced in 1934 by Leray [64] in his quest to understand fluid flow phenomena. Every successful mathematical theory has at its heart a calculation. The calculation behind the Leray theory is the energy inequality. If u, p is a smooth solution to (8.1.1), then multiplying the momentum equation by u, integrating over Ω, integrating by parts, and integrating in time gives

$$\frac{1}{2}||u(t)||^2 + \int_0^t \nu||\nabla u(t')||^2 dt' = \frac{1}{2}||u_0||^2 + \int_0^t (f(t'), u(t'))dt'.$$

The energy equality has the physical interpretation:

$$\text{kinetic energy}(t) + \text{total energy dissipated over}[0,t] \\ = \text{initial kinetic energy} + \text{total power input}.$$

This energy equality also tells us the the right place to look for a solution involves spaces of functions for which $||v(t)|| < \infty$ uniformly on $[0, T]$ and $\int_0^T ||\nabla v(t')||^2 dt' < \infty$.

Definition 28. $Q := L_0^2(\Omega) = \{q(x) \in L^2(\Omega) : \int_\Omega q(x)dx = 0\}$,

$$\begin{aligned} L^2(0,T; H_0^1(\Omega)) &= \left\{ v(x,t) : [0,T] \to H_0^1(\Omega) : \int_0^T ||\nabla v||^2 \, dt < \infty \right\}, \\ L^\infty(0,T; L^2(\Omega)) &= \{v(x,t) : [0,T] \to L^2(\Omega) \text{: ess sup}_{0<t<T} \, ||v|| < \infty\}, \text{ and} \\ L^2(0,T; L_0^2(\Omega)) &= \left\{ q(x,t) : [0,T] \to L_0^2(\Omega)\text{: } \int_0^T ||q(t)||^2 dt < \infty \right\}. \end{aligned}$$

This definition should be interpreted in the usual sense: each space means (more precisely) the closure of smooth functions in the indicated norm. We think, however, of $L^2(0, T; H_0^1(\Omega))$ as consisting of all functions, vanishing on $\partial\Omega$ with

$$\int_0^T ||\nabla v||^2 \, dt < \infty$$

and $L^\infty(0, T; L^2(\Omega))$ as all functions with

$$\sup_{0<t<T} ||v|| < \infty.$$

Thus, $L^\infty(0, T; L^2(\Omega))$ consists of all velocity fields with bounded kinetic energy and $L^2(0, T; H_0^1(\Omega))$ of all velocity fields which dissipate a finite amount of energy in total.

It is important in generalizing a notion of solution that any sufficiently smooth, classical solution of (8.1.1) (should one exist) also be a weak solution. To this end, let us assume (u, p) is a sufficiently smooth solution of (8.1.1) and develop a variational formulation of (8.1.1). Recalling from the steady state case, the natural function spaces for the spatial part of (8.1.1) are

$$\begin{aligned} X :&= \{v \in L^2(\Omega)^d : \nabla v \in L^2(\Omega)^{d\times d} \text{ and } v = 0 \text{ on } \partial\Omega\}, \\ Q :&= \left\{ q \in L^2(\Omega) : \int_\Omega q \, dx = 0 \right\}, \quad \text{and} \\ V :&= \{v \in X : (q, \nabla \cdot v) = 0 \; \forall \, q \in Q\}. \end{aligned}$$

Let (u, p) be a sufficiently smooth solution of (8.1.1). Multiplication of (8.1.1) by $v \in X$ and $q \in Q$ and integration over Ω gives, after applying the divergence theorem,

$$\int_\Omega u_t \cdot v \, dx + \int_\Omega u \cdot \nabla u \cdot v - p \, \nabla \cdot v + \nu \, \nabla u : \nabla v \, dx = \int_\Omega f \cdot v \, dx,$$
$$\int_\Omega q \, \nabla \cdot u \, dx = 0.$$

The integrals above all make sense for any $v \in X, q \in Q$ provided $u \in X$, $p \in Q$ and $u_t \in X^*$.

Motivated by this, we can define one notion of a generalized solution to (8.1.1), a *strong solution,*[61] as follows.

[61]This is not the weakest definition possible of a strong solution.

Definition 29 (strong solution of the NSE). *(u, p) is a* strong solution *of* (8.1.1) *if*

$$u \in L^2(0, T; X) \cap L^\infty(0, T; L^2(\Omega))$$

and

(i) *$u : [0, T] \to X$ is a differentiable map with $u_t \in L^2(0, T; X^*)$ and $p : (0, T] \to Q$ is an integrable map with $p \in L^2(0, T; Q)$.*

(ii) *For every $t' \in (0, T]$, (u, p) satisfies*

$$\int_0^{t'} (u_t, v) + (u \cdot \nabla u, v) - (p, \nabla \cdot v) + \nu(\nabla u, \nabla v)dt = \int_0^{t'} (f, v)dt$$

for all $v \in L^2(0, T; H_0^1(\Omega)) \cap L^\infty(0, T; L^2(\Omega))$ and

$$\int_0^{t'} (q, \nabla \cdot u)dt = 0$$

for all $q \in L^2(0, T; L_0^2(\Omega))$.

(iii) *$u_0 \in V$ and*

$$||u(t) - u_0|| \to 0 \text{ as } t \to 0.$$

(iv) *Additionally, $u \in L^4(0, T; X)$.*

Note that (ii) and the regularity assumed in the definition of a strong solution imply that (except possibly on a set of measure zero) for every $t \in (0, T]$, (u, p) satisfies

$$(u_t, v) + (u \cdot \nabla u, v) - (p, \nabla \cdot v) + \nu(\nabla u, \nabla v) = (f, v).$$

Generally, if u is a strong solution, then formal manipulations like multiplying the equation by u and integrating over the flow domain can be safely performed.

8.2 Weak Solution of the NSE

The notion of weak solution[62] of (8.1.1) is due to Leray [64], who called them *turbulent solutions*. The idea and theory is both mathematically elegant and balanced and physically lucid. To develop this idea, which requires as little regularity as possible on the solution (u, p), first note that space and time integration shows that a solution of (8.1) satisfies

$$(u(T), \phi(T)) + \int_0^T -\left(u, \frac{\partial \phi}{\partial t}\right) + \nu(\nabla u, \nabla \phi) + (u \cdot \nabla u, \phi)\, dt$$
$$= \int_0^T (f, \phi)\, dt + (u(0), \phi(0))$$

[62] In fact, the concept of weak solution of a partial differential equation was introduced in this remarkable paper under the name turbulent solution. The whole theory of distributions and weak solutions of partial differential equations is a direct outgrowth of Leray's development of weak solutions to study the connection between fluid flow phenomena and the behavior of solutions of the Navier–Stokes equations.

for any function $\phi(x,t)$ which is smooth enough, vanishes on $\partial\Omega$, and satisfies

$$\nabla \cdot \phi = 0.$$

To give the precise definition of weak solution will require the introduction of more mathematical structure. In particular, recall the definition of the *support* of a function ϕ in the open set Ω.

Definition 30 (functions with compact support). *The support of ϕ in Ω,* $\text{supp}(\phi)$, *is*

$$\text{supp}(\phi) = \textit{closure}\,\{x \in \Omega : \phi(x) \neq 0\}.$$

A function ϕ has compact support in Ω if $\text{supp}(\phi)$ *is a closed and bounded (i.e., compact) subset of Ω (hence strictly inside Ω).*

Define (following Galdi [38], where the full details are beautifully presented)

$$\mathcal{D}(\Omega) := \{\psi \in C^\infty(\Omega)^d : \psi \text{ has compact support in } \Omega, \text{ and } \nabla\cdot\psi = 0 \text{ in } \Omega\},$$
$$H(\Omega) := \text{ completion of } \mathcal{D}(\Omega) \text{ in } L^2(\Omega)^d, \textit{ and}$$
$$\mathcal{D}_T := \{\phi(x,t) \in C^\infty(\Omega\times[0,T]) : \phi(x,t) \in \mathcal{D}(\Omega) \text{ for each } t,\ 0 \le t \le T\}.$$

The space $H(\Omega)$ is a very interesting function space in its own right. For example, the following is known (see Chapter III of Galdi [37]).

Lemma 21. *The space $H(\Omega)$ can be characterized as follows:*

$$H(\Omega) = \{v \in L^2(\Omega)^d : \nabla\cdot v = 0 \textit{ and } v\cdot\hat{n} = 0 \textit{ on } \Gamma\}.$$

Definition 31 (function spaces for weak solutions).

$$L^2(0,T;V) = \left\{v(t) : [0,T] \to V : \int_0^T ||\nabla v||^2\,dt < \infty\right\} \textit{ and}$$
$$L^\infty(0,T;H) = \{v(t) : [0,T] \to H(\Omega) : \text{ess}\sup_{0<t<T} ||v|| < \infty\}.$$

This definition, like the first one in this chapter, should be interpreted in the usual sense. Each space means (more precisely) the closure of smooth functions (in V or H, respectively) in the indicated norm. Also, as usual, we think of $L^2(0,T;V)$ as consisting of *all* div-free functions, vanishing on $\partial\Omega$ with $\int_0^T ||\nabla v||^2\,dt < \infty$ and $L^\infty(0,T;H)$ as *all* div-free functions with normal component vanishing on $\partial\Omega$ and with $\sup_{0<t<T} ||v|| < \infty$.

Definition 32 (weak solution of the NSE). *Let* $u_0 \in H(\Omega)$, $f \in L^2(\Omega\times(0,T))$. *A measurable function* $u(x,t) : \Omega\times[0,T] \to \mathbb{R}^d$ *is a weak solution of* (8.1) *if*

(i) $u \in L^2(0,T;V) \cap L^\infty(0,T,H(\Omega))$, *and*

(ii) u *satisfies the integral relation* (2.1)*:*

$$(u(T),\phi(T)) + \int_0^T -\left(u, \frac{\partial\phi}{\partial t}\right) + \nu(\nabla u, \nabla\phi) + (u\cdot\nabla u, \phi)\,dt$$
$$= \int_0^T (f,\phi)\,dt + (u(0),\phi(0))$$

for all $\phi \in \mathcal{D}_T$.

(iii) *(Leray's inequality/the energy inequality)*[63] *for any* $t \in [0, T]$

$$\frac{1}{2}||u(t)||^2 + \nu \int_0^t ||\nabla u(t')||^2 dt' \leq \frac{1}{2}||u_0||^2 + \int_0^t (u(t'), f(t'))dt',$$

(iv) $\lim_{t\to\infty} ||u(t) - u_0|| = 0.$

The theory of weak solutions of the NSE is a beautiful one. In two dimensions, it is known that a weak solution exists and is unique. In three dimensions, it is known that weak solutions exist but it is not known if weak solutions are unique. On the other hand, in three dimensions there are slightly more restrictive concepts of solutions to (8.1) which are known to be unique but for which existence is unknown. Resolving this mathematical gap between existence and uniqueness of weak solutions for the three dimension NSE is often referred to as the *fundamental problem* in the mathematical analysis of the NSE and one of the *Clay prize problems* [32]. Incidentally, it is also open if the gap reflects an essential property of real fluids or an inadequacy in the model or analysis. (We will return to this Leray conjecture later.)

So as not to end this section on too murky a tone, we will end by stating the three dimensional existence result (see Galdi [39] for a proof), and one uniqueness partial result. The uniqueness result uses the sharpened bound on the nonlinear term, introduced earlier and valid in two or three dimensions:

$$(u \cdot \nabla v, w) \leq C\sqrt{||u||||\nabla u||}||\nabla v||||\nabla w|| \; \forall u, v, w \in X. \tag{8.2.1}$$

Theorem 21 (existence of weak solutions). *Let* $\Omega \subset \mathbb{R}^d$ *be a domain,* $d = 2$ *or* 3, *and* $T > 0$. *Then, for any data*

$$u_0 \in H(\Omega), \; f \in L^2(\Omega \times (0, T)),$$

there exists at least one weak solution to (8.1.1).

Remark 9 (energy equality versus energy inequality). *Strong solutions satisfy the energy* equality; *Galerkin approximations to weak solutions satisfy the energy* equality. *However, it is not known if the energy* inequality *is the best possible result for weak solutions. In other words, it is not known if there exists a weak solution for which strict inequality holds at some instance in time. This gap is one of the mysteries of the NSE.*

Proposition 15 (uniqueness of strong solutions). *Let* $\Omega \subset \mathbb{R}^d$, $d = 2, 3$. *The NSE have at most one strong solution* u *(with* $\nabla u \in L^4(0, T; L^2(\Omega))$ *). Uniqueness also holds if a weak solution satisfies*

$$u \in L^r(0, T; L^s(\Omega))$$

for some r *and* s *with*

$$\frac{d}{s} + \frac{2}{r} = 1, \; d \leq s \leq \infty.$$

[63]Sometimes a solution satisfying this definition except the energy inequality is called a *mild solution*. It is not known if weak solutions exist which satisfy the energy *equality*.

Proof. We give the proof in the case of a strong solution. Recall that if u is a strong solution in the sense of section 8.1, then, in particular,

$$\int_0^T ||\nabla u||^4(t)\, dt < \infty.$$

Recall the key estimate (8.2.1) on the nonlinear term. Suppose $u = u_2$ and u_1 is another solution. Call $\phi = u_1 - u_2$. Subtraction and the estimate (8.2.1) on the trilinear form gives

$$\frac{1}{2}\frac{d}{dt}||\phi||^2 + \nu\, ||\nabla\phi||^2 = -b(\phi, u_2, \phi) \le C\, ||\phi||^{1/2}\, ||\nabla\phi||^{3/2}\, ||\nabla u_2||. \tag{8.2.2}$$

Next, use the inequality

$$ab \le \frac{\epsilon}{p}a^p + \frac{\epsilon^{-q/p}}{q}b^q, \quad \frac{1}{p} + \frac{1}{q} = 1$$

with $p = 4, q = \frac{4}{3}$, giving

$$(C\, ||\nabla u_2||\, ||\phi||^{1/2})\, (||\nabla\phi||^{3/2}) \le \epsilon(||\nabla\phi||^{3/2})^{4/3} + C(\epsilon)||\nabla u_2||^4\, ||\phi||^2.$$

Picking $\epsilon = \nu$ and inserting this into (8.2.2) gives

$$\frac{d}{dt}\, ||\phi||^2 \le (C||\nabla u_2||^4)\, ||\phi||^2 \text{ or } y' \le a(t)y(t),\ y(0) = 0,$$

where $y(t) = ||\phi||^2(t), a(t) = C||\nabla u_2||^4(t)$. Note $a(t) \in L^1(0, T) : \int_0^T |a(t)|dt < \infty$ (since $||\nabla u|| \in L^4(0, T)$) and $y \ge 0$. Using an integrating factor (which is equivalent to using Gronwall's inequality) implies $y(t) \equiv ||\phi(t)||^2 \equiv 0$. Thus, $u_1 \equiv u_2$. □

Remark 10 (strong solutions and uniqueness). *A weak solution that is sufficiently regular to ensure uniqueness is often called a "strong solution" in the literature. Thus, a weak solution satisfying the condition*

$$u \in L^r(0, T; L^s(\Omega)) \text{ for some } r \text{ and } s \text{ with } \frac{d}{s} + \frac{2}{r} = 1,\ d \le s \le \infty,$$

would also be called a strong solution by many. Uniqueness of weak solutions under this condition was first proven by Leray in 1934 *(briefly at the beginning of section* 22 *of* [64]*). It is best known due to two fundamental papers of Prodi in* 1959 *and Serrin in* 1963 *and is thus called variously Serrin's condition, the Prodi–Serrin condition, or, most correctly, the Leray–Prodi–Serrin condition. This condition is best (as of* 2006*) in the sense that there are no subsequent uniqueness results implying uniqueness under the above condition with* $= 1$ *replaced by* > 1*. The condition we use,* $\nabla u \in L^4(0, T; L^2(\Omega))$*, implies the above condition is thus a stronger assumption than the Leray–Prodi–Serrin condition. On the other hand, the condition* $\nabla u \in L^4(0, T; L^2(\Omega))$ *is also a natural condition (rather than a weakest possible condition). This condition was also proven by Leray in* 1934 *but is best known because of its occurrence in Ladyzhenskaya's book* [62].

If a solution is a strong solution in any of the various definitions used in the literature, we can safely think of multiplying the NSE by u and then integrating. Without the extra regularity the arguments become more delicate. Typically, an estimate sought is derived for a finite dimensional approximation u_n of the weak solution u, and then an estimate for u is recovered (with good technique and good luck) by letting $n \to \infty$.

Remark 11 (how to win the Clay prize (continued)). *The uniqueness equation from the last proof reads*

$$\frac{1}{2}\frac{d}{dt}||\phi||^2 + \nu\, ||\nabla\phi||^2 = -b(\phi, u_2, \phi) = -\int_\Omega \phi \cdot \nabla^s u \cdot \phi dx.$$

Since

$$trace(\nabla^s u) = \nabla \cdot u = 0,$$

$\nabla^s u$ has both positive and negative eigenvalues. Thus, for uniqueness, all that is necessary is to show that in some useful sense the combination of viscosity and the positive eigenvalues of the deformation tensor are not overwhelmed by the growth caused by the negative eigenvalues of the deformation tensor in the sense of resulting in faster than exponential growth.

8.3 Kinetic Energy and Energy Dissipation

> *One can also verify that the total kinetic energy of the liquid remains bounded but it does not seem possible to deduce from this fact that the motion itself remains regular.*
>
> J. Leray, in [64].

> *The* uniqueness *of the flow and* the continuity *of its kinetic energy at the moments of time where it is not regular* are two open problems*; they are related.*
>
> J. Leray, 1954, in [65].

The energy inequality describes the evolution of the kinetic energy in a fluid's flow. Looking at the energy inequality, three terms are involved:

$$\text{the kinetic energy: } k(t) := \frac{1}{|\Omega|}\frac{\rho_0}{2}\int_\Omega |u(x,t)|^2\, dx,$$

$$\text{the energy dissipation rate: } \varepsilon(t) := \frac{\mu}{|\Omega|}\int_\Omega |\nabla u(x,t)|^2\, dx,$$

$$\text{power input via body force flow interaction: } \frac{1}{|\Omega|}\int_\Omega f \cdot u\, dx,$$

where $|\Omega|$ is the volume of the flow domain Ω, ρ_0 is the density, and μ is the viscosity. Note that (as is usually done) the obvious dependence on the size of the flow domain is scaled out of the definitions by dividing by $|\Omega|$. As noted before, in words, the energy estimate states

$$\text{kinetic energy}(t) + \text{total energy dissipated over } (0,t) \le \text{kinetic energy}(0) + \text{total energy input over } (0,t).$$

This is equivalent to

$$k(t) + \int_0^t \varepsilon(t)\, dt \le k(0) + \int_0^t \frac{1}{|\Omega|} \int_\Omega f \cdot u \, dx dt.$$

The á priori bounds which come from the energy inequality tell us that

1. the kinetic energy is uniformly bounded:

$$\sup_{0 \le t \le T} |k(t)| \le C_1(u_0, T, \nu, f) < \infty;$$

2. the total energy dissipated is finite so the energy dissipation rate is integrable:

$$\int_0^T \varepsilon(t)\, dt \le C(u_0, T, \nu, f) < \infty.$$

It is *not* known if the energy inequality holds with inequality replaced by equality. It is *not* known if the energy dissipation rate $\varepsilon(t)$ is bounded for weak solutions in $3d$.[64] If $\varepsilon(t)$ is unbounded, then $\varepsilon(t)$ being integrable gives important information about how fast $\varepsilon(t)$ can blow up to infinity at its singular points. Singular points are not possible to see in a numerical simulation (any computed approximation to the energy dissipation rate must be finite), but often the general trend is seen of $\varepsilon(t)$ being irregular and becoming more so as the meshwidth and time step are reduced. This is depicted in Figure 8.1.

The *Leray conjecture* is that the lack of a uniqueness proof for weak solutions in three dimensions is not due to a weakness of mathematical techniques of the moment but is rather an essential feature. The mathematical realization in the NSE of turbulence in fluid motion is precisely a breakdown of uniqueness of weak solutions. With this in mind, clearly a critical open question in the NSE is how fast the energy dissipation rate $\varepsilon(t)$ can blow up at its singular points:

$$\textit{for what } p \textit{ is } \int_0^T \varepsilon(t)^p \, dt < \infty?$$

The uniqueness theorem for strong solutions of section 8.2 shows that if

$$\int_0^T \varepsilon(t)^p \, dt < \infty \text{ for } p \ge 2,$$

then weak solutions are unique. In the Leray conjecture, this uniqueness case is associated with laminar flow:

$$\text{laminar flow: } \int_0^T \varepsilon(t)^2 \, dt < \infty,$$

$$\text{turbulent flow: } \int_0^T \varepsilon(t)^p \, dt = \infty \text{ for } 1 \le p \le ?? < 2.$$

This conjecture is one of several interesting ones concerning turbulence and the NSEs. A competing conjecture is the *regularity conjecture* that all solutions to the NSE are smooth

[64] It is known that in two dimensions $\varepsilon(t)$ is bounded: $\varepsilon(t) \in L^\infty(0, T)$.

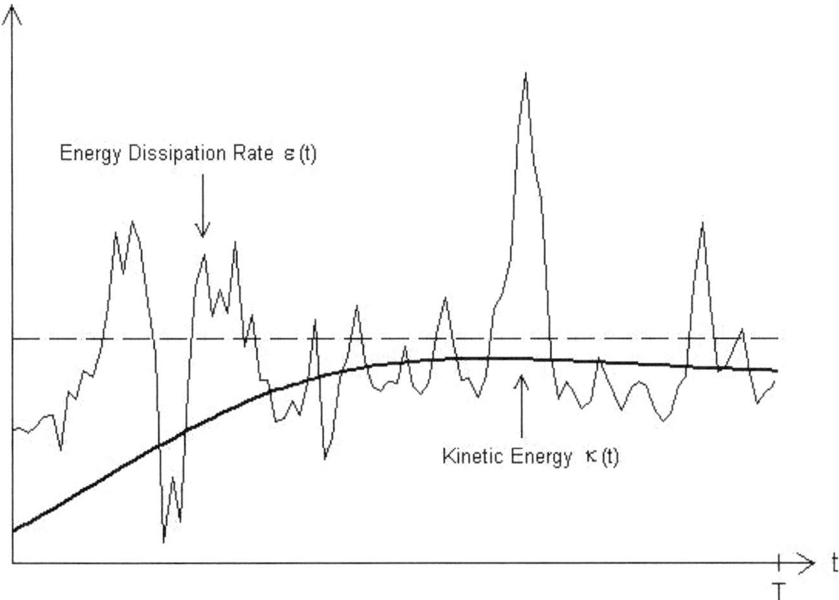

Figure 8.1. *Depiction of k and epsilon.*

strong solutions and the gap between weak and strong solutions is not a physical one but one of only mathematical technique. Turbulence then is not irregularity but rather a multitude of scales from very large to very small but still nonzero. In the regularity conjecture, turbulence has so many scales that there is no effective scale separation and the wave numbers in a turbulent flow are so close together as to appear continuous. The sentiment in the Navier–Stokes community has fluctuated over the years between these two points of view and will likely continue to fluctuate until the gap between weak and strong solutions is closed in the mathematical theory.[65]

8.4 Remarks on Chapter 8

The mathematical theory of the time-dependent NSE is beautiful in those aspects that are complete and fascinating in those aspects that are still unknown. The wonderful article by Galdi [39] is an excellent place to begin reading more about this mathematical theory. The book by Doering and Gibbon [26] is good as well. The condition that is sufficient for uniqueness, that $\varepsilon(t) \in L^2(0, T)$, is originally due to Leray (at the end of section 21 in [64]) and is best known from the exposition in Ladyzhenskaya [62]. It has the advantage that its physical meaning is easy to understand. As noted earlier, it is a stronger assumption than the Leray–Prodi–Serrin condition [64, beginning of section 22], [77], [87]. Of course, the

[65] Another point of view: The Leray conjecture would suggest that solutions of the NSE would contain scales down to zero. The regularity conjecture suggests that the smallest scale, although very small, is bounded away from zero. The latter is more consistent with the assumption that the media is continuous.

fundamental theoretical work on the time-dependent NSE occurred in the landmark paper of Leray [64].

The precise definition of a strong solution varies from one presentation to another. Generally, though, a strong solution is a weak solution which satisfies enough extra regularity (e.g., (iv) in our definition) to ensure that Gronwall's inequality, in some form, can be used to bound differences in velocities. It also implies that some form of the equations like (ii) above holds in which one can set $v = u$. There is also an (obvious but so far unproved) possibility that turbulent velocities are not unique but that the functionals or statistics of turbulent flow which we need are unique.

In the proof of uniqueness of strong solutions, the differential inequality for $y(t) \geq 0$ arose:

$$y'(t) \leq a(t)y(t)\ ,\ y(0) = y_0.$$

If, instead of using an integrating factor, this is integrated, it becomes

$$y(t) \leq y_0 + \int_0^t a(t')y(t')dt',$$

and Gronwall's inequality can be applied. There are many useful generalizations of Gronwall's inequality. In both approaches, the fundamental requirement is that $a(t) \in L^1(0, T)$.

Strong solutions satisfy the energy equality while the best that is known about weak solutions is that they satisfy the energy inequality. This gap between the energy equality and inequality is another mystery of the NSE. Specifically, if there exists a solution and time t^* for which strict inequality holds,

$$\frac{1}{2}||u(t^*)||^2 + \nu \int_0^{t^*} ||\nabla u(t')||^2 dt' < \frac{1}{2}||u_0||^2 + \int_0^{t^*} (u(t'), f(t'))dt',$$

then energy has been dissipated by some mechanism other than molecular dissipation at $t = t^*$. The only possible mechanism is nonlinearity. If this occurs, could it be related to the phenomenon of turbulent bursts? Could it be that equality holds but our mathematical technique isn't good enough yet to prove it? Could it reflect a shortcoming the Navier–Stokes model of fluid motion? None of these possibilities have been ruled out, and this is one of the mysteries of the NSE.

8.5 Exercises

Exercise 95. *Write the momentum equation as*

$$\rho_0(u_t + u \cdot \nabla u) = f + \nabla \cdot \Pi,$$

where Π *is the stress tensor,* $\Pi = \Pi(u, p)$.

(a) *Develop the idea of strong solution with this formulation of the NSE in* (u, p, Π) *variables.*

(b) *Repeat part* (a) *in* (u, p, Π) *variables. You will need to identify an appropriate function space for space for* u, p, *and* Π.

Exercise 96. *Develop a definition of a strong solution in V instead of (X, Q).*

Exercise 97 (uniqueness in two dimensions). *If $\Omega \subset \mathbb{R}^2$ we have the improved estimate*

$$(\phi \cdot \nabla u, \phi) \leq C\ ||\phi||\ ||\nabla\phi||\ ||\nabla u||.$$

Use this estimate to prove that solutions to the NSE are unique in two dimensions.

Exercise 98 (Gronwall's inequality and uniqueness). *Do a library search. Find and present three generalizations of Gronwall's inequality. Pick one and apply it to try to prove uniqueness of solutions to the NSE. Find the extra regularity needed to apply the chosen generalization of Gronwall's inequality.*

Exercise 99 (investigate the Leray–Prodi–Serrin condition). *Beginning with the energy inequality for weak solutions, show that*

$$u \in L^\infty(0, T; L^2(\Omega)) \cap L^2(0, T; H^1(\Omega)).$$

Take $r = s = 4$ in the Leray–Prodi–Serrin condition. Use this regularity, Hölder's inequality, and (6.5.1) *from Chapter* 6 *to show that the Leray–Prodi–Serrin condition holds in two dimensions and verify that the same inequalities fail to verify the Leray–Prodi–Serrin condition in three dimensions.*

Exercise 100 (NSE plus time relaxation and regularity). *Consider the NSE with an added time relaxation term. For $\varepsilon > 0$ small and $\mu \geq 2$,*

$$u_t + u \cdot \nabla u - \nu\Delta u + \varepsilon|u|^{\mu-2}u + \nabla p = f \text{ and } \nabla \cdot u = 0.$$

Prove an energy inequality for this equation. For what values of μ does its solution satisfy the Leray–Prodi–Serrin condition?

Exercise 101 (the time-averaged energy dissipation rate). *Using the energy inequality, find a bound on the time average $\langle\varepsilon\rangle$ of the energy dissipation rate in terms of the time average of the body force, where*

$$\langle\varepsilon\rangle := \limsup_{T\to\infty} \frac{1}{T}\int_0^T \varepsilon(t)\, dt.$$

The bound should be in terms of a reference velocity U and body force F given by

$$U := \sqrt{\left\langle \frac{1}{|\Omega|}||u||^2 \right\rangle} \text{ and } F := \sqrt{\left\langle \frac{1}{|\Omega|}||f||^2 \right\rangle}.$$

Chapter 9
Approximating Time-Dependent Flows

The success of the theory lies not only in the accumulation of elegant mathematical results, but also in its impact on practical computations.
M. Gunzburger, from the preface to [46].

9.1 Introduction

The time dependent NSE are given by

$$u : \Omega \times [0, T] \rightarrow \mathbb{R}^d, \ \ p : \Omega \times (0, T] \rightarrow \mathbb{R}$$

satisfying

$$\begin{aligned} & u_t + u \cdot \nabla u - \nu \Delta u + \nabla p = f \text{ for } x \in \Omega, 0 < t \leq T, \\ & \nabla \cdot u = 0, \ x \in \Omega \text{ for } 0 < t \leq T, \\ & u = 0 \text{ on } \partial\Omega \text{ for } 0 < t \leq T, \\ & u(x, 0) = u_0(x) \text{ for } x \in \Omega \end{aligned} \tag{9.1.1}$$

and the usual normalization condition that $\int_\Omega p(x, t)\, dx = 0$ for $0 < t \leq T$.

For moderate Reynolds number (and bounded time intervals $0 \leq t \leq T$) it is usual to approximate the solution of (9.1.1) by using the usual FEM in Ω, reducing (9.1.1) to a system of ODEs in time, and a time-stepping method for the time variable (i.e., for the system of ODEs). Even moderate Reynolds number flows can exhibit complex behavior (e.g., Figure 9.1). For higher Reynolds numbers, special upwind schemes or special stabilization methods are typically used for the convection term in (9.1.1), while still higher Reynolds numbers lead to the question of turbulence. The mayhem of turbulence leads to special models and methods.

This chapter considers the solution of (9.1.1) at moderate Reynolds numbers, high enough to expect interesting, time-dependent phenomena but not turbulence. It is usual to break the analysis into two steps: semidiscrete, meaning discrete space and continuous time, and fully discrete, meaning both time and space discretized. In both cases, the fundamental

Figure 9.1. *Streamtubes in a "simple" three-dimensional flow.*

step is the stability bound on the approximate velocity. We consider this stability bound at length for both cases. We also give a convergence proof that is essentially an elaboration of the proof of the stability bound applied to the error equation. This error analysis needs Gronwall's inequality and thus (in three dimensions), an extra regularity assumption is needed, such as

$$\nabla u \in L^4(0,T;L^2(\Omega)), \text{ i.e., } \int_0^T ||\nabla u(t')||^4\, dt' < \infty,$$

or (as is more commonly seen)

$$\nabla u \in L^\infty(0,T;L^2(\Omega)), \text{ i.e., ess} \sup_{0\le t\le T} ||\nabla u(\cdot,t)|| < \infty. \tag{9.1.2}$$

To introduce the ideas, we will first consider a time-stepping method. The most stable time-stepping scheme for a system of ODE,

$$y'(t) + f(t, y(t)) = 0, \; y(0) \text{ given},$$

is the backward Euler method: choose $\Delta t > 0$ and let $y_n \cong y(t_n)$, where $t_n = n\Delta t$. The approximations are computed via

$$\frac{y_{n+1} - y_n}{\Delta t} + f(t_{n+1}, y_{n+1}) = 0, \; y_0 \text{ given}.$$

If this scheme is applied to (9.1.1) it gives the following steady problem at each time-step. Given (u_n, p_n) find (u_{n+1}, p_{n+1}) by solving

$$\begin{aligned}
&-\nu\Delta u_{n+1} + (\Delta t)^{-1}u_{n+1} + u_{n+1}\cdot\nabla u_{n+1} + \nabla p_{n+1} \\
&\qquad = f(t_{n+1}, x) + (\Delta t)^{-1}u_n \text{ in } \Omega, \\
&\nabla\cdot u_{n+1} = 0 \text{ in } \Omega, \\
&u_{n+1} = 0 \text{ on } \partial\Omega \text{ and } \int_\Omega p_{n+1}\, dx = 0.
\end{aligned}$$

This leads to solving a sequence of problems which look like the steady NSEs. Each problem's solution begins with data (u_n) with inherent errors caused by the spacial discretization. Thus, one central question is to quantify the inherent errors caused by the spacial discretization. Another central question is how these errors are magnified by the (discrete) time evolution. On the other hand, there are also more accurate time-stepping schemes than backward Euler. Thus, it is customary to divide the error analysis into two steps and begin (as we do in the next section) by studying only the spacial discretization's effect on the error in the time dependent problem. The steps are as follows:

(i) *Semidiscrete approximation.* The spacial variables are discretized by finite elements while the time is held continuous. The evolution of the spacial errors in time is studied.

(ii) *Fully discrete approximation.* The time variable is then discretized by a time-stepping scheme with attractive stability properties. The interplay between spacial errors and time-stepping errors is studied.

We will follow this path and begin by studying the semidiscrete approximation. Recall that the spacial part of (9.1.1) is naturally formulated in

$$X := H_0^1(\Omega)^d, \quad Q := L_0^2(\Omega).$$

Definition 33 (strong solution). *(u, p) is a* strong solution *of* (9.1.1) *if $u \in L^2(0, T; X) \cap L^\infty(0, T; L^2(\Omega))$ and*

(i) *$u : [0, T] \to X$ is a differentiable map with $u_t : (0, T] \to X^*$ an integrable map and $p : (0, T] \to X$ is a continuous map.*

(ii) *For all $t' \in (0, T]$, (u, p) satisfies*

$$\int_0^{t'} (u_t, v) + (u \cdot \nabla u, v) - (p, \nabla \cdot v) + \nu(\nabla u, \nabla v)dt' = \int_0^{t'} (f, v)dt'$$

for all $v \in L^2(0, T; H_0^1(\Omega)) \cap L^\infty(0, T; L^2(\Omega))$ and

$$\int_0^{t'} (q, \nabla \cdot u)dt' = 0$$

for all $q \in L^2(0, T; L_0^2(\Omega))$.

(iii) *$u_0 \in V$ and*

$$||u(t) - u_0|| \to 0 \text{ as } t \to 0.$$

(iv) *$u \in L^4(0, T; X)$.*

In particular, if u is a strong solution, then for a.e. $t' \in (0, T]$, (u, p) satisfies

$$(u_t, v) + (u \cdot \nabla u, v) - (p, \nabla \cdot v) + \nu(\nabla u, \nabla v) = (f, v) \text{ and } (q, \nabla \cdot u) = 0$$

for all $v \in L^2(0, T; H_0^1(\Omega)) \cap L^\infty(0, T; L^2(\Omega))$ and all $q \in L^2(0, T; L_0^2(\Omega))$.

The semidiscrete or continuous-in-time finite element approximation begins by selecting finite element spaces $X^h \subset X$, $Q^h \subset Q$ satisfying condition (LBB^h). The approximate velocity and pressure are maps

$$u^h : [0, T] \to X^h, \ p^h : (0, T] \to Q^h$$

satisfying $\forall\, v^h \in X^h$ and $\forall\, q^h \in Q^h$

$$\begin{aligned}&(u_t^h, v^h) + b^*(u^h, u^h, v^h) + \nu(\nabla u^h, \nabla v^h) - (p^h, \nabla \cdot v^h) = (f, v^h),\\&(\nabla \cdot u^h, q^h) = 0, \ \text{ and } (u^h(\cdot, 0) - u_0, v^h) = 0.\end{aligned} \tag{9.1.3}$$

If we pick a basis for (X^h, Q^h) and expand (u^h, p^h) in terms of that basis, (9.1.3) reduces to a nonlinear system of ODEs in t with a linear side condition. Ultimately, a time-stepping method must also be used to compute approximate velocities and pressures.

9.2 Stability and Convergence of the Semidiscrete Approximations

[A] scheme of weather prediction which resembles the process by which the Nautical Almanac is produced in so far as it is founded upon the differential equations and not upon the partial recurrence of phenomena in their ensemble.
L. F. Richardson, in [81].

It is not altogether obvious if a solution (u^h, p^h) to (9.1.3) exists. To show this, we will eliminate the constraint and reduce (9.1.3) to a system of ordinary differential equations whose solution, we will show, cannot blow up in finite time. Under (LBB^h) the pressure can be eliminated (temporarily) from the system by restricting v^h in (9.1.3) to the space of discretely divergence free functions V^h :

$$V^h := \{v^h \in X^h : (q^h, \nabla \cdot v^h) = 0 \,\forall\, q^h \in Q^h\}.$$

From (9.1.3), $u^h(t) : [0, T] \to V^h$ satisfies

$$(u_t^h, v^h) + b^*(u^h, u^h, v^h) + \nu(\nabla u^h, v^h) = (f, v^h) \ \forall\, v^h \in V^h. \tag{9.2.1}$$

Proposition 16 (stability of the semidiscrete approximation). *The solution to* (9.2.1), u^h, *is stable. For any* $t > 0$

$$\frac{1}{2}||u^h(t)||^2 + \frac{\nu}{2}\int_0^t ||\nabla u^h(t')||^2\,dt' \le \frac{1}{2}||u_0^h||^2 + \frac{\nu^{-1}}{2}\int_0^t ||f(t')||_{-1}^2\,dt'.$$

Proof. Exercise! $\square$

Corollary 6. *The semidiscrete approximation* u^h *exists and is unique.*

Proof. The nonlinearity in the system of ODEs defining u^h is quadratic (and thus locally Lipschitz). Thus, the theory of ordinary differential equations implies that a global-in-time solution exists and is unique as long as every possible solution cannot blow up in finite time. The previous proposition shows exactly that! □

For the convergence proof, we follow the proven plan of decomposing the error into two parts:

$$\begin{aligned} e(t) = u(\cdot,t) - u^h(\cdot,t) &= (u(\cdot,t) - \tilde{u}(\cdot,t)) - (u^h(\cdot,t) - \tilde{u}(\cdot,t)) \\ &= \eta(\cdot,t) - \phi^h(\cdot,t), \end{aligned}$$

where $\tilde{u}(x,t)$ approximates $u(\cdot,t)$ in V^h, $\eta = u - \tilde{u}$, and $\phi(\cdot,t) = (u^h,(\cdot,t) - \tilde{u}(\cdot,t)) \in V^h$. We will derive an energy inequality for ϕ^h and bound ϕ^h in terms of η. The proof will require the sharper bound on the nonlinearity which was the key to proving uniqueness of strong solutions. This upper bound is improvable in two dimensions.

Lemma 22 (a sharper bound on the nonlinear term). *Let $\Omega \subset \mathbb{R}^3$ or $\mathbb{R}^2$. For all $u, v, w \in X$,*

$$b^*(u,v,w) \le C(\Omega)\sqrt{||u||\,||\nabla u||}\,||\nabla v||\,||\nabla w||.$$

Theorem 22 (convergence for strong solutions). *Let u be a strong solution of the NSE (in particular $||\nabla u|| \in L^4(0,T)$). Then, there is a $C^*(T, f, u_0, \nu, \int_0^T ||\nabla u||^4\,dt)$ such that*

$$\begin{aligned} \sup_{0\le t\le T} ||u - u^h||^2 + \nu\int_0^T ||\nabla(u-u^h)||^2\,dt \le C^*\Bigg\{ ||u_0 - u^h(0)||^2 \\ + \inf_{\substack{v^h(t)\in X^h \\ q^h(t)\in Q^h}} \Bigg[\int_0^T ||p - q^h||^2 + ||(u-v^h)_t||_{-1}^2 + \nu||\nabla(u-v^h)||^2 dt \\ + ||\nabla(u-v^h)||_{L^4(0,T;L^2)}^2 + \max_{0\le t\le T} ||u - v^h||^2 \Bigg]\Bigg\}. \end{aligned}$$

Proof. u^h satisfies (9.2.1) so we need a similar equation for u. Note that $V^h \not\subset V$, so extra terms will occur. Multiplying (9.1.1) by $v^h \in V^h$ gives

$$(u_t, v^h) + b^*(u,u,v^h) + \nu(\nabla u, \nabla v^h) + (p, \nabla\cdot v^h) = (f, v^h).$$

Subtracting (9.1.3) from the above, we obtain

$$(e_t, v^h) + [b^*(u,u,v^h) - b^*(u^h,u^h,v^h)] + \nu(\nabla e, \nabla v^h) + (p, \nabla\cdot v^h) = 0 \;\forall\, v^h \in V^h.$$

Let $\widetilde{u} : [0,T] \to V^h$ be arbitrary. Splitting the equation via $e = \eta - \phi^h$, where $\eta = u - \widetilde{u}$, $\phi^h = u^h - \widetilde{u}$, yields

$$\begin{aligned} &(\phi_t^h, v^h) + \nu(\nabla\phi^h, \nabla v^h) = (\eta_t, v^h) + \nu(\nabla\eta, \nabla v^h) \\ &- (p, \nabla\cdot v^h) - [b^*(u,u,v^h) - b^*(u^h,u^h,v^h)]. \end{aligned}$$

Setting $v^h = \phi^h (\in V^h)$ and using standard inequalities gives for any $\epsilon > 0$,

$$\begin{aligned} &\frac{1}{2}\frac{d}{dt}||\phi^h||^2 + \frac{1}{2}(\nu - \epsilon)||\nabla\phi^h||^2 \le \frac{1}{2\epsilon}(||\eta_t||_{-1}^2 + \nu||\nabla\eta||^2) \qquad (9.2.2) \\ &+ (p, \nabla\cdot\phi^h) - [b^*(u,u,\phi^h) - b^*(u^h,u^h,\phi^h)]. \end{aligned}$$

First, note that since $\phi^h \in V^h$, $(q^h, \nabla \cdot \phi^h) = 0$. Thus

$$(p, \nabla \cdot \phi^h) = (p - q^h, \nabla \cdot \phi^h) \; \forall \, q^h \in Q^h.$$

Consider the quadratic term. Using the same idea as in the proof of uniqueness of strong solutions in Chapter 8, we decompose the nonlinearity as

$$\begin{aligned} &b^*(u, u, \phi^h) - b^*(u^h, u, \phi^h) + b^*(u^h, u, \phi^h) - b^*(u^h, u^h, \phi^h) \\ &= b^*(e, u, \phi^h) + b^*(u^h, e, \phi^h) = b^*(\eta, u, \phi^h) - b^*(\phi^h, u, \phi^h) + b^*(u^h, \eta, \phi^h). \end{aligned}$$

Using these two identities in (9.2.2) yields

$$\begin{aligned} &\frac{1}{2}\frac{d}{dt}||\phi^h||^2 + \frac{1}{2}(\nu - \epsilon)||\nabla \phi^h||^2 \leq \frac{1}{2\epsilon}(||\eta_t||_{-1}^2 + \nu||\nabla \eta||^2) \\ &+ (p - q^h, \nabla \cdot \phi^h) + b^*(\eta, u, \phi^h) + b^*(u^h, \eta, \phi^h) - b^*(\phi^h, u, \phi^h). \end{aligned} \tag{9.2.3}$$

The last three, nonlinear terms on the right-hand side of (9.2.3) require careful treatment. The first term can be bounded simply by

$$b^*(\eta, u, \phi^h) \leq M||\nabla \eta|| \, ||\nabla u|| \, ||\nabla \phi^h|| \leq \epsilon ||\nabla \phi^h||^2 + \frac{1}{4\epsilon} M^2 ||\nabla u||^2 \, ||\nabla \eta||^2.$$

The third nonlinear term is really the key to the error analysis and needs a bound a bit sharper than the above. Lemma 22 yields the upper bound:

$$b^*(\phi^h, u, \phi^h) \leq C||\phi^h||^{1/2}||\nabla \phi^h||^{3/2}||\nabla u||.$$

The idea is to "hide" the $||\nabla \phi^h||$ term and take care of the $||\phi^h||$ term using Gronwall's inequality. For this the usual inequality $ab \leq \frac{\epsilon}{2} a^2 + \frac{1}{2\epsilon} b^2$ is not sufficient and we use instead the following interesting inequality from Exercise 1 in Chapter 1: given $a, b > 0$, for any $\epsilon > 0$, $1 \leq p \leq \infty$, $\frac{1}{p} + \frac{1}{q} = 1$,

$$ab \leq \epsilon a^p + C(\epsilon, p) \, b^q.$$

With $p = 4/3$, $p' = 4$ we obtain

$$\begin{aligned} b^*(\phi^h, u, \phi^h) &\leq ||\nabla \phi^h||^{3/2}(C||\nabla u|| \, ||\phi^h||^{1/2}) \\ &\leq \epsilon ||\nabla \phi^h||^2 + C(\epsilon) \, ||\nabla u||^4 \, ||\phi^h||^2. \end{aligned}$$

Using this bound in (9.2.3) gives

$$\begin{aligned} &\frac{1}{2}\frac{d}{dt}||\phi^h||^2 + \frac{1}{2}(\nu - 5\epsilon)||\nabla \phi^h||^2 \leq \frac{1}{2\epsilon}(||\eta_t||_{-1}^2 + \nu||\nabla \eta||^2) \\ &+ (p - q^h, \nabla \cdot \phi^h) + C(\epsilon)||\nabla u||^4||\phi^h||^2 + C(\epsilon)||\nabla u||^2||\nabla \eta||^2 \\ &+ b^*(u^h, \eta, \phi^h). \end{aligned} \tag{9.2.4}$$

Note that $(p - q^h, \nabla \cdot \phi^h) \leq \epsilon ||\nabla \phi^h||^2 + C\epsilon^{-1}||p - q^h||^2$. Using Lemma 22 on the last nonlinear term $b^*(u^h, \eta, \phi^h)$ yields

$$\begin{aligned} b^*(u^h, \eta, \phi^h) &\leq C||u^h||^{1/2}||\nabla u^h||^{1/2}||\nabla \eta|| \, ||\nabla \phi^h|| \\ &\leq \epsilon ||\nabla \phi^h||^2 + C(\epsilon)||u^h|| \, ||\nabla u^h|| \, ||\nabla \eta||^2. \end{aligned}$$

By the stability bound

$$\sup_{0\le t\le T} ||u^h|| + ||\nabla u^h||_{L^2(0,T;L^2)} \le C(f, u_0, \nu),$$

we have

$$b^*(u^h, \eta, \phi^h) \le \epsilon||\nabla\phi^h||^2 + C(\epsilon, f, u_0)||\nabla u^h||\,||\nabla\eta||^2.$$

Using these inequalities in (9.2.3) we obtain

$$\frac{1}{2}\frac{d}{dt}||\phi^h||^2 + \frac{1}{2}(\nu - 9\epsilon)||\nabla\phi^h||^2 \le C\epsilon^{-1}||p - q^h||^2$$
$$+ C\epsilon^{-1}[||\eta_t||_{-1}^2 + \nu||\nabla\eta||^2] + C(\epsilon)||\nabla u||^2\,||\nabla\eta||^2 + C(\epsilon, f, u_0, \nu)||\nabla u^h||\,||\nabla\eta||^2$$
$$- C(\epsilon)||\nabla u||^4\,||\phi^h||^2.$$

Picking $\epsilon = \nu/18$ gives

$$\frac{1}{2}\frac{d}{dt}||\phi^h||^2 + \frac{1}{4}\nu||\nabla\phi^h||^2 \le C(\nu)[||p - q^h||^2 + ||\eta_t||_{-1}^2 + \nu^{-1}||\nabla\eta||^2$$
$$+ ||\nabla u||^2\,||\nabla\eta||^2 + ||\nabla u^h||\,||\nabla\eta||^2] + C(\mathrm{Re})||\nabla u||^4\,||\phi^h||^2.$$

Let $a(t) := C(\nu)||\nabla u||^4$. If $a(t) \in L^1(0, T)$, then we can form its antiderivative

$$A(t) := \int_0^t a(t')dt' < \infty \text{ for } ||\nabla u|| \in L^4(0, T).$$

Multiplying through by the integrating factor $e^{-A(t)}$ gives

$$\frac{d}{dt}[e^{-A(t)}||\phi^h||^2] + e^{-A(t)}\nu||\nabla\phi^h||^2$$
$$\le e^{-A(t)}C(\nu)[||p - q^h||^2 + ||\eta_t||_{-1}^2$$
$$+ \nu^{-1}||\nabla\eta||^2 + ||\nabla u||^2\,||\nabla\eta||^2 + ||\nabla u^h||\,||\nabla\eta||^2].$$

Integrating this over $[0, t]$ and multiplying through by $e^{A(t)}$ gives a bound of ϕ^h in terms of η of the form[66]

$$||\phi^h(t)|| + \nu\int_0^t ||\nabla\phi^h||^2 dt'$$
$$\le C^*(T, \nu)||\phi^h(0)||^2 + C^*(T, \nu)\int_0^t \Big[||\eta_t||_{-1}^2 + ||\nabla\eta||^2$$
$$+ ||\nabla u||^2||\nabla\eta||^2 + ||\nabla u^h||||\nabla\eta||^2 + \inf_{q^h\in Q^h}||p - q^h||^2\Big]\,dt'.$$

The one problematic term is the last one involving η. Applying Hölder's inequality gives

$$\int_0^t ||\nabla u^h(t')||\,||\nabla\eta(t')||^2 dt' \le ||\nabla u^h||_{L^2(0,T;L^2)}||\nabla\eta||_{L^4(0,T;L^2)}^2.$$

[66]These last few steps are equivalent to using Gronwall's inequality.

Note that $||\nabla u^h||_{L^2(0,T;L^2)}$ is bounded by problem data by the stability bound.

Using this bound and the triangle inequality completes the proof. □

Remark 12 (variations on the error analysis). *The fundamental error equation is (from the last proof) ($e = \eta - \phi^h$, where $\eta = u - \widetilde{u}$, $\phi^h = u^h - \widetilde{u}$):*

$$(\phi_t^h, v^h) + \nu(\nabla\phi^h, \nabla v^h) = (\eta_t, v^h) + \nu(\nabla\eta, \nabla v^h)$$
$$- (p - q^h, \nabla \cdot \phi^h) - [b^*(u, u, v^h) - b^*(u^h, u^h, v^h)].$$

The different types of estimates and different possible paths in the error estimation begin here with the choice of $\widetilde{u}$. The above theorem presents the most basic options for each. If, instead of the interpolant, $\widetilde{u}$ is chosen to be a Stokes projection into V^h, then

$$(\nabla\eta, \nabla v^h) = 0 \;\forall v^h \in V^h.$$

Thus the second term on the right-hand side of this error equation vanishes identically and the error analysis and resulting estimates change. (This requires precise error estimates for the Stokes projection.) Another option is to choose $\widetilde{u}$ as the time-dependent Stokes projection into V^h; then

$$(\eta_t, v^h) + \mathrm{Re}^{-1}(\nabla\eta, \nabla v^h) = 0 \forall v^h \in V^h.$$

Thus both the first and the second term on the right-hand side of this error equation vanishes identically and the error equation becomes

$$(\phi_t^h, v^h) + \nu(\nabla\phi^h, \nabla v^h) = -(p - q^h, \nabla \cdot \phi^h) - [b^*(u, u, v^h) - b^*(u^h, u^h, v^h)].$$

The analysis and resulting estimates change again. (This requires still more precise error estimates for the time-dependent Stokes projection.) With a choice of $\widetilde{u}$ made, the next option changes the left-hand side by the choice of v^h. The basic choice (used above) is $v^h = \phi^h$. This leads to an estimate (in the energy norm) of the error of the form

$$\max_{0 \le t \le T} ||u - u^h||^2 + \nu \int_0^T ||\nabla(u - u^h)||^2 \, dt \le \cdots .$$

The second option is to choose $v^h = t\phi^h$(because it vanishes at $t = 0$). Terms result weighted by t (vanishing at $t = 0$). This option yields error estimates which require less smoothness of the initial data.

The third possible choice is $v^h = \phi_t^h$; this leads to an estimate of the form

$$\int_0^T ||u_t - u_t^h||^2 dt + \max_{0 \le t \le T} ||\nabla(u - u^h)||^2 \le \cdots .$$

The fourth option is the choice of the initial condition for the approximation. This can make more subtle changes depending on the above choices.

9.3 Rates of Convergence

When studying any physical problem in applied mathematics, three essential stages are involved.

(i) Modelling:...

(ii) Mathematical study of the problem:...

(iii) Numerical analysis of the model: By this is meant the description of, and the mathematical analysis of, approximation schemes which can *be run on a computer in "reasonable" time to get "reasonably accurate" numbers.*

P. Ciarlet, in *Lectures on Three Dimensional Elasticity*, Springer, New York, 1983.

The question remains: How fast does u^h converge to u? Do we obtain a useful answer in a useful time (or in geologic time)? This is fundamentally a question about rate of convergence. To evaluate the rates of convergence as $h \to 0$, we must make a specific choice of X^h, Q^h and a specific choice of v^h, q^h in Theorem 22. To begin, we study the interpolant in $X^h = C^0$ piecewise linear vector functions. Let $u(x, t)$ be the fluid velocity and suppose, for now, that $u \in C^1(\Omega \times (0, T))$. Let e_i denote the standard unit vectors in $\mathbb{R}^d$ and let the nodal basis for X^h be denoted by

$$\{\phi_j(x)e_i : \text{all nodes } N_j \text{ inside } \Omega \text{ and } i = 1, \dots, d\}.$$

Then, for each $t \in [0, T]$ the piecewise linear interpolant $I^h(u)$ is given by

$$(I^h u)(x, t) = \sum_{\text{all nodes } N_j \text{ in mesh}} \sum_{l=1}^{d} u_l(N_j, t)\phi_j(x)e_l.$$

The interpolant of $\frac{\partial u}{\partial t}$ is, similarly,

$$I^h\left(\frac{\partial u}{\partial t}\right)(x, t) = \sum_{\text{all nodes } N_j \text{ in mesh}} \sum_{l=1}^{d} \frac{\partial u_l}{\partial t}(N_j, t)\phi_j(x)e_l.$$

Comparing these two formulas gives the following observation.

Lemma 23. *Suppose the nodes N_j in the finite element mesh do not move (i.e., are independent of t). Let $X^h \supset$ (C^0 piecewise linears)$^d \cap X$. Then,*

$$\frac{\partial}{\partial t}(I^h u) = I^h\left(\frac{\partial u}{\partial t}\right) \quad \textit{and}$$

$$\left\|\frac{\partial u}{\partial t} - \frac{\partial}{\partial t} I^h(u)\right\| \le Ch \left\|\nabla \frac{\partial u}{\partial t}\right\| \quad \textit{and}$$

$$\left\|\frac{\partial u}{\partial t} - \frac{\partial}{\partial t} I^h(u)\right\|_{-1} \le Ch \left\|\nabla \frac{\partial u}{\partial t}\right\|.$$

Proof. The first formula follows from the above explicit formulas for the linear interpolant. The second follows from the first and the standard estimate for the error in linear interpolation from Chapter 2. The third follows since $|| \cdot ||_{-1} \le || \cdot ||$. $\square$

In many cases the above estimate in $H^{-1}(\Omega)$ is not sharp and can be improved by a technique from finite element error analysis known as duality; see, e.g., Brenner and

Scott [15]. A typical result when duality can be applied is

$$||v - I^h(v)||_{-1} \leq Ch||v - I^h(v)||,$$

and thus

$$||v - I^h(v)||_{-1} \leq Ch^3||\nabla\nabla v||, \qquad (9.3.1)$$
$$||v - I^h(v)||_{-1} \leq Ch^2||\nabla v||.$$

Using these estimates we can give a bound for the error in the time-dependent problem that predicts a rate of convergence as $h \to 0$ for common choices of finite element spaces. To begin we note the following estimate on the interpolation error.

Lemma 24 (interpolation errors). *Let $X^h_{;} Q^h$ denote the finite element spaces associated with the MINI element and let I^h denote the interpolant in the space of C^0 piecewise linears. Suppose the above interpolation estimates (9.3.1) in $H^{-1}(\Omega)$ hold and that $u, u_t, p \in C^0(\Omega \times [0, T])$.*[67] *Then,*

$$\int_0^T ||p - I^h p||^2 + ||u_t - I^h(u_t)||^2_{-1} + ||\nabla(u - I^h u)||^2 dt$$
$$\leq Ch^2 \int_0^T ||\nabla p||^2 + ||\nabla u_t||^2 + ||\nabla\nabla u||^2 dt$$

and

$$||\nabla(u - I^h u)||^2_{L^4(0,T;L^2(\Omega))} \leq Ch^2||\nabla\nabla u||^2_{L^4(0,T;L^2(\Omega))} \text{ and}$$
$$\sup_{0 \leq t \leq T} ||u - I^h u||^2 \leq Ch^2 \sup_{0 \leq t \leq T} ||\nabla\nabla u||^2.$$

Proof. At each time level $||p(\cdot, t) - I^h p(\cdot, t)||^2 \leq Ch^2||\nabla p(\cdot, t)||^2$. Squaring and integrating this over $[0, T]$ gives the first two terms in the first estimate. The remainder follows analogously. □

Theorem 23 (rates of convergence). *Let X^h, Q^h be the finite element spaces associated with the MINI element and let I^h denote the interpolant in the space of C^0 piecewise linears. Suppose the above interpolation estimates (9.3.1) in $H^{-1}(\Omega)$ hold and that $u, u_t, p \in C^0(\Omega \times [0, T])$. Suppose additionally that the assumptions of Theorem 22 hold. Then,*

$$\sup_{0 \leq t \leq T} ||u - u^h||^2 + \nu \int_0^T ||\nabla u - \nabla u^h||^2 dt$$
$$\leq C^*(T, \nu)h^2 \Big\{ ||\nabla u||^2_{L^\infty(0,T;L^2(\Omega))} + ||\nabla\nabla u||^2_{L^4(0,T;L^2(\Omega))}$$
$$+ h^2||\nabla u_t||^2_{L^2(0,T;L^2(\Omega))} + ||\nabla\nabla u||^2_{L^2(0,T;L^2(\Omega))} + ||p||^2_{L^2(0,T;L^2(\Omega))} \Big\}.$$

[67] Recall that for the usual nodal interpolant to be well defined the functions being interpolated should be continuous. This continuity condition can be relaxed a lot by an important technical extension, due to Clement, of interpolation of local (integral) averages of possibly discontinuous functions, e.g., [15].

Proof. This is a direct combination of the two previous results. □

The Hood–Taylor element (a very good element for finite element CFD) consists of C^0 piecewise quadratics for velocity and C^0 piecewise linears for pressure. If the Hood–Taylor element is used instead of the MINI element, the analogous result is

$$\sup_{0\le t\le T} ||u - u^h||^2 + \mathrm{Re}^{-1}\int_0^T ||\nabla u - \nabla u^h||^2 dt \le C(u, p, \mathrm{Re})h^4.$$

It is interesting to understand how big the right-hand side of these and other error estimates are with respect to the Reynolds number since for many simulations Re is much bigger than the mesh width is small. This is often an essential question (which we shall skip herein)! For moderately high Reynolds numbers and computationally feasible meshes, the answer seems to be that these types of convergence estimates give little useful information on the errors because the right-hand side is $O(1)$ or bigger. Yet, CFD simulations produce many interesting and useful results. There is thus a clear analytical challenge here: *Understand what can be computed accurately and what are the errors in those quantities measured in their appropriate manner when the global error degrades to* $O(1)$.

9.4 Time-Stepping Schemes

Finite arithmetical differences have proven remarkably successful in dealing with differential equations... in this book it is shown that similar methods can be extended to the very complicated system of differential equations which express the changes in the weather.
L. F. Richardson, in [81].

The semidiscrete FEM reduces the NSE to a large, stiff system of ordinary differential equations in time. This system must still be solved by an appropriate time-stepping scheme. The NSE have three features that influence the choice of the methods used for time discretization:

1. the stiffness of the problem in the diffusion-dominated flow regions,

2. the overall sensitivity of the underlying physical problem and lack of regularity of its solution, and

3. the large number of mesh points needed in space that leads to an extremely large system of ODEs.

For these reasons, the preference in practice has been to use low order, implicit methods for time-stepping. One example follows.

ALGORITHM 9.1. Trapezoid Rule.
Given a time-step $k = \Delta t > 0$, the one-leg trapezoid rule computes $u_1^h, u_2^h, \ldots, p_1^h, p_2^h, \ldots,$ where $t_j = jk$ and $u_j^h(x) \cong u(x, t_j)$, $p_j^h(x) \cong p(x, t_j)$. At every time-step the following

nonlinear system is solved. Given $(u_n^h, p_n^h) \in (X^h, Q^h)$ find $(u_{n+1}^h, p_{n+1}^h) \in (X^h, Q^h)$ satisfying

$$\left(\frac{u_{n+1}^h - u_n^h}{k}, v^h\right) + \nu\left(\nabla\left(\frac{u_{n+1}^h + u_n^h}{2}\right), \nabla v^h\right) + b^*\left(\frac{u_{n+1}^h + u_n^h}{2}, \frac{u_{n+1}^h + u_n^h}{2}, v^h\right)$$
$$-\left(\frac{1}{2}\left(p_{n+1}^h + p_n^h\right), \nabla \cdot v^h\right) = (f(t_{n+\frac{1}{2}}), v^h)\ \forall\, v^h \in X^h,$$
$$(\nabla \cdot u_{n+1}^h, q^h) = 0\ \forall\, q^h \in Q^h. \quad (9.4.1)$$

This trapezoidal method is A-stable and second order accurate.[68]

Proposition 17 (stability). *The method* (9.4.1) *is unconditionally stable. For any* $h > 0$ *and any* t_n

$$\frac{1}{2}||u_n^h||^2 + k\sum_{l=0}^{n-1}\frac{\nu}{2}||\nabla\left(\frac{u_{l+1}^h + u_l^h}{2}\right)||^2 \leq \frac{1}{2}||u_0^h||^2 + \frac{\nu^{-1}}{2}k\sum_{l=0}^{n-1}||f(t_{l+\frac{1}{2}})||_{-1}^2.$$

Proof. We set $v^h = \frac{u_{n+1}^h + u_n^h}{2}$ and mimic the proof of the energy inequality. We get

$$\frac{1}{2}||u_{n+1}^h||^2 - \frac{1}{2}||u_n^h||^2 + k\frac{\nu}{2}||\nabla\left(\frac{u_{n+1}^h + u_n^h}{2}\right)||^2 = k\left(f(t_{n+\frac{1}{2}}), \frac{u_{n+1}^h + u_n^h}{2}\right).$$

The remainder of the proof is left as an exercise. (*Hint:* Sum this instead of integrating it.) □

The discrete stability bound and its proof have clear parallels with the continuous time case once one recognizes that

$$k\sum_{l=0}^{n-1}\frac{\nu}{2}||\nabla\left(\frac{u_{l+1}^h + u_l^h}{2}\right)||^2 = \text{Riemann sum of } \int_0^{t_n}\frac{\nu}{2}||\nabla u^h(t)||^2 dt.$$

The general strategy of such an energy proof in the discrete case is to (i) make a choice of test function which makes the nonlinear term vanish, (ii) with this choice identify something like an discrete kinetic energy, E_n^h, so the first term becomes

$$(1 + O(k))E_{n+1}^h - (1 + O(k))E_n^h,$$

and (iii) show that with this choice of test function the viscous term is nonnegative and thus is a discrete energy dissipation rate ε_n^h. After that the difference inequality is summed (instead of integrated) or a discrete Gronwall inequality is applied.

Many variants of (9.4.1) are used in practice, including different methods for different time-steps, different terms, and different interpretations of the nonlinearity. For example,

[68] A-stability is a linear stability theory which is necessary but not sufficient for unconditional stability for nonlinear initial value problems. CFD requires nonlinearity stability which is verified by energy methods.

the *two-leg* trapezoid rule is identical to the above except for the nonlinearity. In the two-leg trapezoid rule the nonlinear term is replaced as follows:

$$b^*\left(\frac{u_{n+1}^h + u_n^h}{2}, \frac{u_{n+1}^h + u_n^h}{2}, v^h\right) \Leftarrow \frac{1}{2}b^*(u_{n+1}^h, u_{n+1}^h, v^h) + \frac{1}{2}b^*(u_n^h, u_n^h, v^h).$$

The two-leg version is more convenient in that the nonlinearity can be split easily. For moderate Reynolds number and well-resolved, laminar flows, computational experience with both suggests that the results are not much different between the two. On the other hand, for higher Reynolds numbers and underresolved flows, it is generally believed the one-leg version has stability properties that outweigh its modest extra inconvenience.[69] Interestingly, a similar conclusion holds for explicit skew-symmetrization of the nonlinearity.

One difficulty of the trapezoid rule is that a nonlinear system must be solved each time-step. If (after expanding the left-hand side) we replace

$$b^*(u_{n+1}^h, u_{n+1}^h, v^h) \Leftarrow b^*(u_n^h, u_{n+1}^h, v^h),$$

then one needs to solve only one linear system per time-step, but the accuracy is reduced to only $O(\Delta t)$. This method is commonly used in CFD but is only first order accurate. There is a very good, second order, and unconditionally stable variant of it, the extrapolated trapezoid rule, that requires only one linear system per time-step. To motivate the extrapolated Trapezoid rule, note that if $u(t)$ is smooth, then linear extrapolation gives

$$u(t_{n+\frac{1}{2}}) = \frac{3}{2}u(t_n) - \frac{1}{2}u(t_{n-1}) + O(k^2).$$

Define

$$E[u_n^h, u_{n-1}^h] := \frac{3}{2}u_n^h - \frac{1}{2}u_{n-1}^h.$$

Thus if we replace the nonlinear term with the above extrapolation of it, second order accuracy is preserved:

$$b^*((u_{n+1}^h + u_n^h)/2, (u_{n+1}^h + u_n^h)/2, v^h) \Leftarrow b^*(E[u_n^h, u_{n-1}^h], (u_{n+1}^h + u_n^h)/2, v^h).$$

ALGORITHM 9.2. TRLE: Trapezoid Rule + Linear Extrapolation.
Given a time step $k = \Delta t > 0$, the TRLE method computes $u_1^h, u_2^h, \dots, p_1^h, p_2^h, \dots,$ where $t_j = jk$ and $u_j^h(x) \cong u(x, t_j)$, $p_j^h(x) \cong p(x, t_j)$. At every time step $n \geq 1$ the following linear system is solved. Given $(u_n^h, p_n^h) \in (X^h, Q^h)$ find $(u_{n+1}^h, p_{n+1}^h) \in (X^h, Q^h)$ satisfying

$$\left(\frac{u_{n+1}^h - u_n^h}{k}, v^h\right) + \nu\left(\nabla\left(\frac{u_{n+1}^h + u_n^h}{2}\right), \nabla v^h\right) + b^*\left(E[u_n^h, u_{n-1}^h], \frac{u_{n+1}^h + u_n^h}{2}, v^h\right)$$
$$- \left(\frac{1}{2}(p_{n+1}^h + p_n^h), \nabla \cdot v^h\right) = (f(t_{n+\frac{1}{2}}), v^h) \;\forall\, v^h \in X^h,$$
$$(\nabla \cdot u_{n+1}^h, q^h) = 0 \;\forall\, q^h \in Q^h. \quad (9.4.2)$$

[69]A deeper look at the two- versus one-leg question in [34] makes this difference smaller: averages of the two-leg trapezoid rule satisfy the one-leg trapezoid rule! This means that the complexity of the one-leg version is reduced by ordering the calculation properly and the stability of the two-leg version is enhanced by averaging.

Note that once the first step in the extrapolated trapezoid rule is taken, it requires only one linear system per time-step thereafter. This method is unconditionally stable and second order accurate. The analysis of this method was pioneered in a widely circulated but unpublished report by Baker [9]. It has been further developed in the work of Simo and Armero [5], [89] and He [47].

All implicit methods require at least the solution of a linear system per time-step. Due to the large number of degrees of unknowns in CFD, iterative methods are usually used, e.g., Elman, Silvester, and Wathen [31]. In fact, *when CFD practitioners say that a "method fails," this usually means that the chosen linear or nonlinear iterative solver failed to converge at some time step.* At moderately high Reynolds numbers and moderate Δt this type of failure happens all too often, irrespective of any mathematical proof that the discrete equations have a solution. There is a real need for simple discretizations that are highly accurate and unconditionally stable and which lead to *linear and nonlinear systems that are easier to solve* by known methods *without increasing the energy dissipation in the scheme.*

The subject of stabilizations in CFD is a very active research area. It began with simple ideas (which are computationally inexpensive and don't work so well) like artificial viscosity and upwinding and has advanced to some methods based on weighted local residuals (which are not so simple but work well). The latter are important but beyond this first treatment so we shall give a few intermediate examples. These are simple in motivation, are easy to understand mathematically, and work better than artificial viscosity (at least). As a first example, consider the following simple method, which decreases the Reynolds number for the nonlinear system for the new time level and corrects for it at the old time level. Unfortunately, it is only first order accurate.

ALGORITHM 9.3. BECE-stab: Backward Euler + Constant Extrapolation, Stabilized. Given a time-step $k = \Delta t > 0$, the BECE method computes $u_1^h, u_2^h, \dots, p_1^h, p_2^h, \dots,$ where $t_j = jk$ and $u_j^h(x) \cong u(x, t_j)$, $p_j^h(x) \cong p(x, t_j)$. At every time-step $n \geq 0$ the following linear system is solved. Given $(u_n^h, p_n^h) \in (X^h, Q^h)$ find $(u_{n+1}^h, p_{n+1}^h) \in (X^h, Q^h)$ satisfying

$$\begin{aligned} \left(\frac{u_{n+1}^h - u_n^h}{k}, v^h\right) + (\nu + h)(\nabla u_{n+1}^h, \nabla v^h) + b^*(u_n^h, u_{n+1}^h, v^h) \\ - (p_{n+1}^h, \nabla \cdot v^h) = (f(t_{n+1}), v^h) + h(\nabla u_n^h, \nabla v^h) \ \forall\, v^h \in X^h, \\ (\nabla \cdot u_{n+1}^h, q^h) = 0 \ \forall\, q^h \in Q^h. \end{aligned} \tag{9.4.3}$$

The extra stabilization terms added are

$$h(\nabla(u_{n+1}^h - u_n^h), \nabla v^h) = hk\left(\nabla\left(\frac{u_{n+1}^h - u_n^h}{k}\right), \nabla v^h\right) = O(hk).$$

The accuracy is thus determined by the accuracy of the backward Euler method and is then $O(k + hk)$. This stabilized method is unconditionally stable but only first order accurate in time (due to being based on backward Euler). Each step requires the solution of one linear Oseen problem at Reynolds number no larger than $O(h^{-1})$. This reduced Reynolds number often makes fast solvers more robust, e.g., [31]. Variations on this method have been known for a long time in CFD.[70] Some recent refinements are given in [2].

There are many possible attempts to generalize the stabilization to higher order time-stepping without much success. The one exception seems to be the simplest extension which actually works reasonably well, e.g., [63].

ALGORITHM 9.4. TRLE-stab: Trapezoid Rule + Linear Extrapolation, Stabilized. Given a time-step $k = \Delta t > 0$, the TRLEstab method computes $u_1^h, u_2^h, \ldots, p_1^h, p_2^h, \ldots,$ where $t_j = jk$ and $u_j^h(x) \cong u(x, t_j)$, $p_j^h(x) \cong p(x, t_j)$. At every time-step $n \geq 1$ the following linear system is solved. Given $(u_n^h, p_n^h) \in (X^h, Q^h)$ find $(u_{n+1}^h, p_{n+1}^h) \in (X^h, Q^h)$ satisfying

$$\left(\frac{u_{n+1}^h - u_n^h}{k}, v^h\right) + \nu\left(\nabla\left(\frac{u_{n+1}^h + u_n^h}{2}\right), \nabla v^h\right)$$

$$+ \frac{\alpha}{2}h(\nabla u_{n+1}^h, \nabla v^h) + b^*\left(E[u_n^h, u_{n-1}^h], \frac{u_{n+1}^h + u_n^h}{2}, v^h\right)$$

$$-\left(\frac{1}{2}(p_{n+1}^h + p_n^h), \nabla \cdot v^h\right) = (f(t_{n+\frac{1}{2}}), v^h) + \frac{\alpha}{2}h(\nabla u_n^h, \nabla v^h) \;\forall\, v^h \in X^h,$$

$$(\nabla \cdot u_{n+1}^h, q^h) = 0 \;\forall\, q^h \in Q^h.$$

Note that once the first step is taken, this requires only one linear system per time-step. This method is unconditionally stable and second order, $O(k^2 + hk)$, accurate.

9.5 Convergence Analysis of the Trapezoid Rule

This section gives a classical and basic error analysis of the fully discrete, trapezoid FEM approximation of the NSEs. The proof given herein follows the first in the field which used numerical ODE ideas and were based on a discrete version of the energy estimate. Discrete time methods lead to their own technicalities, notations, and techniques (which mirror the continuous time case). The usual trapezoid rule is a good method and one for which the steps are well understood. We will first present the notation and definitions necessary for the analysis of its convergence.

9.5.1 Notation for the discrete time method

For functions $v(x, t)$ and $1 \leq p \leq \infty$ define

$$\|v\|_{\infty,k} := \operatorname{ess}\sup_{0<t<T} \|v(t,\cdot)\|_k \quad \text{and} \quad \|v\|_{p,k} := \left(\int_0^T \|v(t,\cdot)\|_k^p \, dt\right)^{1/p}.$$

[70]The first time I saw it described was in 1980 and it seemed to have been known for some time then.

We assume the velocity-pressure spaces are conforming, satisfy the LBB^h condition and the following approximation properties:

$$\inf_{v\in X^h} \|u - v\| \le Ch^{k+1}\|u\|_{k+1}, \quad u \in H^{k+1}(\Omega)^d,$$
$$\inf_{v\in X^h} \|u - v\|_1 \le Ch^{k}\|u\|_{k+1}, \quad u \in H^{k+1}(\Omega)^d,$$
$$\inf_{r\in Q^h} \|p - r\| \le Ch^{s+1}\|p\|_{s+1}, \quad p \in H^{s+1}(\Omega).$$

Hood–Taylor elements are one common example of (X^h, Q^h).

We will also need to have notation for discrete time analogues of the above. We denote $v(t_{n+1/2}) = v((t_n + t_{n+1})/2)$ if $v(t)$ is a continuous function of t. The averages of two time levels are denoted $v_{n+1/2} = (v_n + v_{n+1})/2$ for both continuous and discrete functions of t.

We will need the usual bounds on the trilinear form. At some point we shall shorten the analysis a bit by assuming higher regularity of the velocity than usual. This can be improved to parallel the semidiscrete error analysis.

Lemma 25 (continuity). *For $u, v, w \in X$, $b^*(u, v, w)$ satisfies*

$$b^*(u, v, w) \le M(\Omega)\, \|\nabla u\|\, \|\nabla v\|\, \|\nabla w\|\,,$$
$$b^*(u, v, w) \le C(\Omega)\, \|u\|^{\frac{1}{2}}\, \|\nabla u\|^{\frac{1}{2}}\, \|\nabla v\|\, \|\nabla w\|\,,$$

while for $u \in L^2(\Omega)$, $w \in X$, and $v, \nabla v \in L^\infty(\Omega)$

$$b^*(u, v, w) \le \frac{1}{2}\{\|u\|\, \|\nabla v\|_\infty\, \|w\| + \|u\|\, \|v\|_\infty\, \|\nabla w\|\}.$$

Since (X^h, Q^h) satisfies the LBB^h condition, the trapezoid rule is equivalent to

$$\left(\frac{u^h_{n+1} - u^h_n}{\Delta t}, v^h\right) + b^*(u^h_{n+1/2}, u^h_{n+1/2}, v^h)$$
$$+ \nu(\nabla u^h_{n+1/2}, \nabla v^h) = (f_{n+1/2}, v^h)\ \forall\ v^h \in V^h.$$

The next lemma will be used to bound the consistency error of the trapezoid rule.

Lemma 26 (consistency errors). *Let Δt be a fixed positive real number. If $u = u(t)$ is smooth enough, then*

$$\left\| u_t\left(t + \frac{\Delta t}{2}\right) - \frac{u(t+\Delta t) - u(t)}{\Delta t} \right\| \le C\Delta t^2\, \|u_{ttt}\|_{L^\infty(t,t+\Delta t; L^2(\Omega))}\,,$$
$$\left\| u_{n+1/2} - u(t_{n+1/2}) \right\|^2 \le \frac{1}{48}(\Delta t)^3 \int_{t_n}^{t_{n+1}} \|u_{tt}\|^2\, dt\,,$$
$$\left\| \frac{u(t_{n+1}) - u(t_n)}{\Delta t} - u_t(t_{n+1/2}) \right\|^2 \le \frac{1}{1280}(\Delta t)^3 \int_{t_n}^{t_{n+1}} \|u_{ttt}\|^2\, dt\,, \text{ and}$$
$$\left\| \nabla(u_{n+1/2} - u(t_{n+1/2})) \right\|^2 \le \frac{(\Delta t)^3}{48} \int_{t_n}^{t_{n+1}} \|\nabla u_{tt}\|^2\, dt\,.$$

Proof. The proof is simply Taylor expansion with integral remainder. Expand $u(t)$ and $u(t+\Delta t)$ about $u(t+\frac{\Delta t}{2})$; use the integral remainder; square both sides and integrate. $\square$

In the continuous time approximation, the final step in the error analysis is to either use an integrating factor or equivalently integrate and apply Gronwall's lemma. The discrete time case is analogous. However, difference equations are more complex than differential equations and there are many discrete Gronwall inequalities. We give the next two versions, which are commonly used in discrete time error analysis. The first discrete Gronwall inequality (see [51]) is as follows.

Lemma 27 (a discrete Gronwall lemma). *Let Δt, H, and a_n, b_n, c_n, d_n (for integers $n \geq 0$) be nonnegative numbers such that*

$$a_l + \Delta t \sum_{n=0}^{l} b_n \leq \Delta t \sum_{n=0}^{l} d_n a_n + \Delta t \sum_{n=0}^{l} c_n + H \textit{ for } l \geq 0.$$

Suppose that $\Delta t d_n < 1 \ \forall n$. Then,

$$a_l + \Delta t \sum_{n=0}^{l} b_n \leq exp\left(\Delta t \sum_{n=0}^{l} \frac{d_n}{1-\Delta t d_n}\right)\left(\Delta t \sum_{n=0}^{l} c_n + H\right) \quad \textit{for } l \geq 0.$$

This version has the smallness condition: $\Delta t d_n < 1$. Sometimes this condition is innocuous and sometimes it is restrictive. If the first sum on the right-hand side stops before the sum on the left-hand side, it can be dropped (e.g., a remark to Lemma 5.1 in [51]).

Lemma 28 (another discrete Gronwall lemma). *Let $\Delta t, B, a_n, b_n, c_n, d_n$ for integers $n \geq 0$ be nonnegative numbers such that for $l \geq 1$. If*

$$a_l + \Delta t \sum_{n=0}^{l} b_n \leq \Delta t \sum_{n=0}^{l-1} d_n a_n + \Delta t \sum_{n=0}^{l} c_n + B \ \textit{ for } l \geq 0,$$

then for all $\Delta t > 0$,

$$a_l + \Delta t \sum_{n=0}^{l} b_n \leq \exp(\Delta t \sum_{n=0}^{l-1} d_n)\left(\Delta t \sum_{n=0}^{l} c_n + B\right) \quad \textit{for } l \geq 0.$$

The discrete time error analysis requires norms that are discrete time analogues of the norms used in the continuous time case:

$$|||v|||_{\infty,k} := \max_{0\leq n\leq N_T} \|v^n\|_k, \qquad |||v_{1/2}|||_{\infty,k} := \max_{1\leq n\leq N_T} \|v^{n-1/2}\|_k,$$

$$|||v|||_{p,k} := \left(\sum_{n=0}^{N_T} \|v^n\|_k^p \Delta t\right)^{1/p}, \qquad |||v_{1/2}|||_{p,k} := \left(\sum_{n=1}^{N_T} \|v^{n-1/2}\|_k^p \Delta t\right)^{1/p}.$$

9.5.2 Error analysis of the trapezoid rule

We show that solutions of the fully discrete, trapezoid rule in time FEM in space scheme are unconditionally stable, well defined, and optimally convergent to solutions of the NSE. This error analysis is notationally technical but the ideas and steps are parallels to the continuous time error analysis. To keep the volume of inequalities reasonable, we shall at several steps use inequalities that are a bit less sharp that the ones used in the continuous time error analysis.

Lemma 29 (stability). *Consider the fully discrete trapezoidal FEM approximation. A solution u_h^l, $l = 1, \dots M$, exists at each time-step. The scheme is also unconditionally stable. It satisfies the following á priori bound:*

$$\left\|u_M^h\right\|^2 + \nu\Delta t \sum_{n=0}^{M-1} \left\|\nabla u_{n+1/2}^h\right\|^2 \le \left\|u_0^h\right\|^2 + \frac{\Delta t}{\nu}\sum_{n=0}^{M-1} \left\|f_{n+1/2}\right\|_*^2.$$

Proof. For existence, note that given u_n^h, the method requires solution of a nonlinear system for u_{n+1}^h. This system is very similar to the one arising from the FEM approximation to the steady state NSE. Existence of a solution u_{n+1}^h is proved in the same way as in the steady state case (an exercise in applying the Leray–Schauder fixed point theorem).

To obtain the á priori estimate, set $v^h = u_{n+1/2}^h$ in the discrete equations. After applying the Cauchy–Schwarz and Young inequalities, this gives

$$\frac{1}{\Delta t}\left(\left\|u_{n+1}^h\right\|^2 - \left\|u_n^h\right\|^2\right) + \nu\left\|\nabla u_{n+1/2}^h\right\|^2 \le \frac{1}{\nu}\left\|f_{n+1/2}\right\|_*^2 \ \forall n.$$

Summing from $n = 0 \dots M-1$ gives the desired result. □

The main convergence result is given next.

Theorem 24. *Let $(u(t), p(t))$ be a sufficiently smooth, strong solution of the NSE* (9.1.1). *Suppose (u_0^h, p_0^h) are approximations of $(u(0), p(0))$ to within the accuracy of the interpolant. Then there is a constant $C = C(u,p)$ such that*

$$|||u - u^h|||_{\infty,0} \le F(\Delta t, h) + Ch^{k+1}|||u|||_{\infty,k+1} \text{ and}$$

$$\left\{\nu\Delta t\sum_{n=0}^{M-1}\|\nabla(u_{n+1/2} - (u_{n+1}^h + u_n^h)/2)\|^2\right\}^{\frac{1}{2}} \le F(\Delta t, h) + C\nu^{1/2}\Delta t^2\|\nabla u_{tt}\|_{2,0} + C\nu^{1/2}h^k|||u|||_{2,k+1},$$

where $F(\Delta t, h) = O(\Delta t^2 + h^k)$:

$$\begin{aligned}F(\Delta t,h) &:= C\nu^{-1/2}h^{k+1/2}\left(|||u|||_{4,k+1}^2 + |||\nabla u|||_{4,0}^2\right)\\ &+ C\nu^{-1/2}h^k\left(|||u|||_{4,k+1}^2 + \nu^{-1/2}(\|u_0^h\| + \nu^{-1/2}|||f|||_{2,\star})\right)\\ &+ C\nu^{-1/2}h^{k+1}|||p_{1/2}|||_{2,k+1} + C\nu^{1/2}h^k|||u|||_{2,k+1}\\ &+ C\Delta t^2\left(\|u_{ttt}\|_{2,0} + \nu^{-1/2}\|p_{tt}\|_{2,0} + \|f_{tt}\|_{2,0} + \nu^{1/2}\|\nabla u_{tt}\|_{2,0}\right.\\ &\left.+ \nu^{-1/2}\|\nabla u_{tt}\|_{4,0}^2 + \nu^{-1/2}|||\nabla u|||_{4,0}^2 + \nu^{-1/2}|||\nabla u_{1/2}|||_{4,0}^2\right).\end{aligned}$$

In the case of Hood–Taylor elements, we have the following.

Corollary 7. *Suppose that in addition to the assumptions of the above theorem, X^h and Q^h are composed of Hood–Taylor elements. Then the error is of the order*

$$|||u - u^h|||_{\infty,0} + \left\{ \nu\Delta t \sum_{n=1}^{M} \|\nabla(u_{n+1/2} - u^h_{n+1/2})\|^2 \right\}^{\frac{1}{2}} = O(h^2 + \Delta t^2).$$

***Proof of Theorem* 9.5.1.** Let U denote the L^2 projection of u into V^h. It will reduce the number of terms in the error analysis by one to take $u^h_0 = U(0)$.

First the equation of the true solution of the NSE is rearranged to be the discrete time trapezoid method plus whatever terms are left over. These leftover terms make up the methods consistency error, τ (evaluated at the true solution), and are estimated by a Taylor expansion.

Then, at time $t_{n+1/2}$, the true solution u satisfies, by rearrangement,

$$\left(\frac{u_{n+1} - u_n}{\Delta t}, v^h\right) + b^*(u_{n+1/2}, u_{n+1/2}, v^h) + \nu(\nabla u_{n+1/2}, \nabla v^h)$$
$$- (p_{n+1/2}, \nabla \cdot v^h) = (f_{n+1/2}, v^h) + \tau(u_n, p_n; v^h)$$

for all $v^h \in V^h$, where $\tau(u_n, p_n; v^h)$ is the consistency error. It is

$$\begin{aligned}\tau(u_n, p_n; v^h) = &\left(\frac{u_{n+1} - u_n}{\Delta t} - u_t(t_{n+1/2}), v^h\right) \\ &+ \nu(\nabla u_{n+1/2} - \nabla u(t_{n+1/2}), \nabla v^h) \\ &+ b^*(u_{n+1/2}, u_{n+1/2}, v^h) - b^*(u(t_{n+1/2}), u(t_{n+1/2}), v^h) \\ &- (p_{n+1/2} - p(t_{n+1/2}), \nabla \cdot v^h) + (f(t_{n+1/2}) - f_{n+1/2}, v^h)\,.\end{aligned}$$

Subtracting the above form of the NSE from the trapezoid FEM equations and letting $e_n = u_n - u^h_n$ gives the basic error equation of the method:

$$\left(\frac{e_{n+1} - e_n}{\Delta t}, v^h\right) + b^*(u_{n+1/2}, u_{n+1/2}, v^h) - b^*(u^h_{n+1/2}, u^h_{n+1/2}, v^h)$$
$$+ \nu(\nabla e_{n+1/2}, \nabla v^h) = (p_{n+1/2}, \nabla \cdot v^h) + \tau(u_n, p_n; v^h) \quad \forall v^h \in V^h.$$

Decompose the error as $e_n = (u_n - U_n) - (u^h_n - U_n) := \eta_n - \phi^h_n$, where $\phi^h_n \in V^h$. Setting $v^h = \phi^h_{n+1/2}$, the above basic error equation equation, using $(q^h, \nabla \cdot \phi_{n+1/2}) = 0$ for all $q^h \in V^h$, can be rewritten as

$$(\phi^h_{n+1} - \phi^h_n, \phi^h_{n+1/2}) + \nu\Delta t\|\nabla\phi^h_{n+1/2}\| + \Delta t\, b^*(u^h_{n+1}, e_{n+1/2}, \phi^h_{n+1/2})$$
$$+ \Delta t\, b^*(e_{n+1/2}, u_{n+1/2}, \phi^h_{n+1/2}) = (\eta_{n+1} - \eta_n, \phi^h_{n+1/2})$$
$$+ \Delta t\nu(\nabla\eta_{n+1/2}, \nabla\phi^h_{n+1/2}) + \Delta t(p_{n+1/2} - q, \nabla \cdot \phi^h_{n+1/2}) + \Delta t\,\tau(u_n, p_n; v^h)\,.$$

Adding and subtracting in the nonlinear terms yields

$$
\begin{aligned}
\frac{1}{2}(\|\phi^h_{n+1}\| - \|\phi^h_n\|) + \nu\Delta t\|\nabla\phi^h_{n+1/2}\| &= (\eta_{n+1} - \eta_n, \phi^h_{n+1/2}) \\
&+ \Delta t \nu(\nabla\eta_{n+1/2}, \nabla\phi^h_{n+1/2}) - \Delta t\, b^*(\eta_{n+1/2}, u_{n+1/2}, \phi^h_{n+1/2}) \\
&+ \Delta t\, b^*(\phi^h_{n+1/2}, u_{n+1/2}, \phi^h_{n+1/2}) - \Delta t\, b^*(u^h_{n+1/2}, \eta_{n+1/2}, \phi^h_{n+1/2}) \\
&+ \Delta t(p_{n+1/2} - q, \nabla\cdot\phi^h_{n+1/2}) + \Delta t\, \tau(u_n, p_n; v^h)\,.
\end{aligned}
$$

We now bound the last right-hand side term by term in the next five inequalities. Recall that U is the L^2 projection of u in V^h so that $(\eta_{n+1} - \eta_n, \phi^h_{n+1/2}) = 0$. The Cauchy–Schwarz and Young inequalities give

$$
\begin{aligned}
\nu\Delta t(\nabla\eta_{n+1/2}, \nabla\phi^h_{n+1/2}) &\le \nu\Delta t\|\nabla\eta_{n+1/2}\|\,\|\nabla\phi^h_{n+1/2}\| \\
&\le \frac{\nu\Delta t}{12}\left\|\nabla\phi^h_{n+1/2}\right\|^2 + C\nu\Delta t\left\|\nabla\eta_{n+1/2}\right\|^2 \quad\text{and} \\
\Delta t(p_{n+1/2} - q, \nabla\cdot\phi^h_{n+1/2}) &\le C\Delta t\|p_{n+1/2} - q\|\,\|\nabla\phi^h_{n+1/2}\| \\
&\le \frac{\nu\Delta t}{12}\left\|\nabla\phi^h_{n+1/2}\right\|^2 + C\Delta t\nu^{-1}\|p_{n+1/2} - q^h\|^2.
\end{aligned}
$$

Standard bounds on the trilinear form and the usual inequalities give

$$
\begin{aligned}
&\Delta t\, b^*(\eta_{n+1/2}, u_{n+1/2}, \phi^h_{n+1/2}) \\
&\le C\Delta t\|\eta_{n+1/2}\|^{1/2}\,\|\nabla\eta_{n+1/2}\|^{1/2}\,\|\nabla u_{n+1/2}\|\,\|\nabla\phi^h_{n+1/2}\| \\
&\le \frac{\nu\Delta t}{12}\|\phi^h_{n+1/2}\|^2 + C\Delta t\,\nu^{-1}\|\eta_{n+1/2}\|\,\|\nabla\eta_{n+1/2}\|\|\nabla u_{n+1/2}\|^2, \\
&\Delta t\, b^*(\phi^h_{n+1/2}, u_{n+1/2}, \phi^h_{n+1/2}) \\
&\le C\Delta t\|\phi^h_{n+1/2}\|^{1/2}\,\|\nabla\phi^h_{n+1/2}\|^{1/2}\,\|\nabla u_{n+1/2}\|\,\|\nabla\phi^h_{n+1/2}\| \\
&\le C\Delta t\|\phi^h_{n+1/2}\|^{1/2}\,\|\nabla\phi^h_{n+1/2}\|^{3/2}\,\|\nabla u_{n+1/2}\| \\
&\le \frac{\nu\Delta t}{12}\|\nabla\phi^h_{n+1/2}\|^2 + C\Delta t\,\nu^{-3}\|\phi^h_{n+1/2}\|^2\,\|\nabla u_{n+1/2}\|^4, \\
&\Delta t\, b^*(u^h_{n+1/2}, \eta_{n+1/2}, \phi^h_{n+1/2}) \\
&\le C\|u^h_{n+1/2}\|^{1/2}\,\|\nabla u^h_{n+1/2}\|^{1/2}\,\|\nabla\eta_{n+1/2}\|\,\|\nabla\phi^h_{n+1/2}\| \\
&\le \frac{\nu\Delta t}{12}\|\nabla\phi^h_{n+1/2}\|^2 + C\Delta t\,\nu^{-1}\|u^h_{n+1/2}\|\,\|\nabla u^h_{n+1/2}\|\,\|\nabla\eta_{n+1/2}\|^2.
\end{aligned}
$$

Combining the bounds on the last five estimates in (103) and summing from $n = 0$ to $M - 1$ (using that $\|\phi_0^h\| = 0$) reduces (5.17) to

$$
\begin{aligned}
&\|\phi_M^h\|^2 + \nu\Delta t \sum_{n=0}^{M-1} \|\nabla\phi_{n+1/2}^h\|^2 \\
&\le \Delta t \sum_{n=0}^{M-1} C\nu^{-3}\|\nabla u_{n+1/2}\|^4 \, \|\phi_{n+1/2}^h\|^2 + \Delta t \sum_{n=0}^{M-1} C\nu\|\nabla\eta_{n+1/2}\|^2 \\
&+\Delta t \sum_{n=0}^{M-1} C\nu^{-1}\|\eta_{n+1/2}\| \, \|\nabla\eta_{n+1/2}\| \|\nabla u_{n+1/2}\|^2 \\
&+\Delta t \sum_{n=0}^{M-1} C\nu^{-1}\|u_{n+1/2}^h\| \, \|\nabla u_{n+1/2}^h\| \, \|\nabla\eta_{n+1/2}\|^2 \\
&+\Delta t \sum_{n=0}^{M-1} C\nu^{-1}\|p_{n+1/2} - q\|^2 + \Delta t \sum_{n=0}^{M-1} C|\tau(u_n, p_n, \phi_{n+1/2}^h)| \,.
\end{aligned}
\tag{9.5.1}
$$

The proof continues by estimating the second, third, and fourth terms on the above right-hand side. The second term is evaluated by the approximation assumption as

$$
\begin{aligned}
\Delta t \sum_{n=0}^{M-1} C\nu\|\nabla\eta_{n+1/2}\|^2 &\le \Delta t C\nu \sum_{n=0}^{M} \|\nabla\eta_n\|^2 \le \Delta t C\nu \sum_{n=0}^{M} h^{2k}|u_n|_{k+1}^2 \\
&\le C\nu h^{2k} \||u|\|_{2,k+1}^2 \,.
\end{aligned}
$$

For the third term, we have

$$
\begin{aligned}
&\Delta t \sum_{n=0}^{M-1} C\nu^{-1}\|\eta_{n+1/2}\| \, \|\nabla\eta_{n+1/2}\| \, \|\nabla u_{n+1/2}\|^2 \\
&\le C\nu^{-1}\Delta t \sum_{n=0}^{M-1} (\|\eta_{n+1}\| \, \|\nabla\eta_{n+1}\| + \|\eta_n\| \, \|\nabla\eta_n\| \\
&\qquad + \|\eta_n\| \, \|\nabla\eta_{n+1}\| + \|\eta_{n+1}\| \, \|\nabla\eta_n\|)\|\nabla u_{n+1/2}\|^2 \\
&\le C\,\nu^{-1}\,h^{2k+1} \left\{ \Delta t \sum_{n=0}^{M-1} |u_{n+1}|_{k+1}^2 \, \|\nabla u_{n+1/2}\|^2 \right. \\
&\quad \left. + \; \Delta t \sum_{n=0}^{M-1} |u_{n+1}|_{k+1}|u_n|_{k+1} \, \|\nabla u_{n+1/2}\|^2 + \Delta t \sum_{n=0}^{M-1} |u_n|_{k+1}^2 \, \|\nabla u_{n+1/2}\|^2 \right\} \\
&\le C\nu^{-1}\,h^{2k+1} \left(\Delta t \sum_{n=0}^{M} |u_n|_{k+1}^4 + \Delta t \sum_{n=0}^{l} \|\nabla u_n\|^4 \right) \\
&\le C\nu^{-1}\,h^{2k+1} \left(\||u|\|_{4,k+1}^4 + \||\nabla u|\|_{4,0}^4 \right) \,.
\end{aligned}
$$

Using the a priori estimate for $\|u_n^h\|$ in the stability lemma yields

$$\Delta t \sum_{n=0}^{M-1} C\nu^{-1} \left(\|u_{n+1/2}^h\| \, \|\nabla u_{n+1/2}^h\| \, \|\nabla \eta_{n+1/2}\|^2\right)$$

$$\leq C\nu^{-1}\Delta t \sum_{n=0}^{M-1} \|\nabla u_{n+1/2}^h\| \, \|\nabla \eta_{n+1/2}\|^2$$

$$\leq C\nu^{-1}\Delta t \sum_{n=0}^{M-1} \left(\|\nabla \eta_{n+1}\|^2 + \|\nabla \eta_n\|^2\right) \|\nabla u_{n+1/2}^h\|$$

$$\leq C\nu^{-1}h^{2k}\Delta t \sum_{n=0}^{M-1} \left(|u_{n+1}|_{k+1}^2 + |u_n|_{k+1}^2\right) \|\nabla u_{n+1/2}^h\|$$

$$\leq C\nu^{-1}h^{2k} \left(\Delta t \sum_{n=0}^{M} |u_n|_{k+1}^4 + \Delta t \sum_{n=0}^{M} \|\nabla u_{n+1/2}^h\|^2\right)$$

$$\leq C\nu^{-1}h^{2k} \left(\|\|u\|\|_{4,k+1}^4 + \nu^{-1}(\|u^h\|^2 + \nu^{-1}\|\|f\|\|_{2,\star}^2)\right) .$$

Adding and subtracting terms, using the Taylor expansion lemma and the approximation assumptions, we bound the pressure term on the right-hand side of (9.5.1) by

$$\Delta t \sum_{n=0}^{M-1} \nu^{-1}\|p_{n+\frac{1}{2}} - q\|^2 \leq C\nu^{-1}\Delta t \sum_{n=0}^{M-1} \{\|p(t_{n+\frac{1}{2}}) - q\|^2 + \|p_{n+\frac{1}{2}} - p(t_{n+\frac{1}{2}})\|^2\}$$

$$\leq C\nu^{-1} \left(h^{2s+2}\Delta t \sum_{n=0}^{M-1} \|p(t_{n+\frac{1}{2}})\|_{s+1}^2 + \Delta t \sum_{n=0}^{M-1} \frac{1}{48}\Delta t^3 \int_{t_n}^{t_{n+1}} \|p_{tt}\|^2 \, dt\right)$$

$$\leq C\nu^{-1} \left(h^{2s+2}\|\|p_{\frac{1}{2}}\|\|_{2,s+1}^2 + (\Delta t)^4 \|p_{tt}\|_{2,0}^2\right) .$$

Next consider the consistency error term on the right-hand side of (9.5.1), $\tau(u_n, p_n; \phi_{n+\frac{1}{2}}^h)$. The consistency error term has many components. All are combinations of the error in a central difference approximation to a time derivative and the error in approximation by averaging. For the first component, using the Cauchy–Schwarz and Young inequalities and the Taylor series lemma, we arrive at the estimates

$$\left(\frac{u_{n+1} - u_n}{\Delta t} - u_t(t_{n+\frac{1}{2}}), \phi_{n+\frac{1}{2}}^h\right)$$

$$\leq \frac{1}{2}\|\phi_{n+\frac{1}{2}}^h\|^2 + \frac{1}{2}\left\|\frac{u_{n+1} - u_n}{\Delta t} - u_t(t_{n+1/2})\right\|^2$$

$$\leq \frac{1}{2}\|\phi_{n+1}^h\|^2 + \frac{1}{2}\|\phi_n^h\|^2 + \frac{1}{2}\frac{(\Delta t)^3}{1280}\int_{t_n}^{t_{n+1}} \|u_{ttt}\|^2 \, dt ,$$

The consistency error term has a contribution from the pressure. It is estimated as

$$
\begin{aligned}
&(p_{n+1/2} - p(t_{n+1/2}), \nabla \cdot \phi^h_{n+1/2}) \\
&\le \varepsilon_1 \nu \|\nabla \phi^h_{n+1/2}\|^2 + C\,\nu^{-1}\|p_{n+1/2} - p(t_{n+1/2})\|^2 \\
&\le \varepsilon_1 \nu \|\nabla \phi^h_{n+1/2}\|^2 + C\,\nu^{-1}\frac{(\Delta t)^3}{48}\int_{t_n}^{t_{n+1}} \|p_{tt}\|^2\,dt\,.
\end{aligned}
$$

The body force term in the consistency error is bounded similarly:

$$
\begin{aligned}
&(f(t_{n+1/2}) - f_{n+1/2}, \phi^h_{n+1/2}) \\
&\le \frac{1}{2}\|\phi^h_{n+1/2}\|^2 + \frac{1}{2}\|f(t_{n+1/2}) - f_{n+1/2}\|^2 \\
&\le \frac{1}{2}\|\phi^h_{n+1}\|^2 + \frac{1}{2}\|\phi^h_{n}\|^2 + \frac{(\Delta t)^3}{48}\int_{t_n}^{t_{n+1}} \|f_{tt}\|^2\,dt\,.
\end{aligned}
$$

By similar estimates we have

$$
\begin{aligned}
&(\nabla u_{n+1/2} - \nabla u(t_{n+1/2}), \nabla \phi^h_{n+1/2}) \\
&\le \varepsilon_2 \nu\, \|\nabla \phi^h_{n+1/2}\|^2 + C\,\nu\|\nabla u_{n+1/2} - \nabla u(t_{n+1/2})\|^2 \\
&\le \varepsilon_2 \nu\, \|\nabla \phi^h_{n+1/2}\|^2 + C\,\nu\frac{\Delta t^3}{48}\int_{t_n}^{t_{n+1/2}} \|\nabla u_{tt}\|^2\,dt\,.
\end{aligned}
$$

The nonlinear term in the consistency error is estimated as follows:

$$
\begin{aligned}
&b^*(u_{n+1/2}, u_{n+1/2}, \phi^h_{n+1/2}) - b^*(u(t_{n+1/2}), u(t_{n+1/2}), \phi^h_{n+\frac{1}{2}}) \\
&= b^*(u_{n+1/2} - u(t_{n+1/2}), u_{n+1/2}, \phi^h_{n+1/2}) - b^*(u(t_{n+1/2}), u_{n+1/2} - u(t_{n+1/2}), \phi^h_{n+1/2}) \\
&\le C\,\|\nabla(u_{n+1/2} - u(t_{n+1/2}))\|\,\|\nabla \phi^h_{n+1/2}\|\,\big(\|\nabla u_{n+1/2}\| + \|\nabla u(t_{n+1/2})\|\big) \\
&\le C\,\nu^{-1}\left(\|\nabla u_{n+1/2}\|^2 + \|\nabla u(t_{n+1/2})\|^2\right)\frac{\Delta t^3}{48}\int_{t_n}^{t_{n+1}} \|\nabla u_{tt}\|^2\,dt + \varepsilon_3\nu\|\nabla \phi^h_{n+1/2}\|^2 \\
&\le C\,\nu^{-1}\Delta t^3\left(\int_{t_n}^{t_{n+1}} \|\nabla u_{n+1/2}\|^4 + \|\nabla u(t_{n+1/2})\|^4 + \|\nabla u_{tt}\|^4\,dt\right) + \varepsilon_3\nu\|\nabla \phi^h_{n+1/2}\|^2 \\
&\le C\,\nu^{-1}\,\Delta t^4(\|\nabla u_{n+1/2}\|^4 + \|\nabla u(t_{n+1/2})\|^4) \\
&\quad + C\,\nu^{-1}\,\Delta t^3\int_{t_n}^{t_{n+1}} \|\nabla u_{tt}\|^4\,dt + \varepsilon_3\nu\|\nabla \phi^h_{n+1/2}\|^2\,.
\end{aligned}
$$

Combining the estimates of the last five terms yields the following upper bound for the consistency error term in (9.5.1):

$$\Delta t \sum_{n=0}^{M-1} |\tau(u_n, p_n; \phi^h_{n+1/2})| \le \Delta t\, C\|\phi^h_{n+1}\|^2 + (\varepsilon_1 + \varepsilon_2 + \varepsilon_3 + \varepsilon_4)\Delta t\, \nu\|\nabla\phi^h_{n+1/2}\|^2$$
$$+ C\nu^{-1}(\delta^2 h^{2k} + h^{2k+2})\left(\sum_{n=1}^{N} \| \, (A^{-1})^N u \, \|^2_{2,k+1}\right)$$
$$+ C(\Delta t)^4 \left\{\|u_{ttt}\|^2_{2,0} + \nu^{-1}\|p_{tt}\|^2_{2,0} + \|f_{tt}\|^2_{2,0}\right.$$
$$+ \nu\|\nabla u_{tt}\|^2_{2,0} + \nu^{-1}\|\nabla u_{tt}\|^4_{4,0}$$
$$\left. + \ \nu^{-1}|||\nabla u|||^4_{4,0} + \nu^{-1}|||\nabla u_{1/2}|||^4_{4,0}\right\}.$$

Inserting all these estimates in (9.5.1), taking $\varepsilon_1 = \varepsilon_2 = \varepsilon_3 = \varepsilon_4 = 1/12$ gives (after collecting terms)

$$\|\phi^h_M\|^2 + \nu\Delta t \sum_{n=0}^{M-1} \|\nabla\phi^h_{n+1/2}\|^2$$
$$\le \Delta t \sum_{n=0}^{M-1} C(\nu^{-3}\|\nabla u_{n+1/2}\|^4 + 1)\, \|\phi^h_{n+1/2}\|^2 + C\nu h^{2k}|||u|||^2_{2,k+1}$$
$$+ C\nu^{-1} h^{2k+1}\left(|||u|||^4_{4,k+1} + |||\nabla u|||^4_{4,0}\right) + C\nu^{-1}h^{2s+2}|||p_{1/2}|||^2_{2,s+1}$$
$$+ C\nu^{-1}h^{2k}\left(|||u|||^4_{4,k+1} + \nu^{-1}\|u^h\|^2 + \nu^{-2}|||f|||^2_{2,\star}\right)$$
$$+ \ C\Delta t^4(\|u_{ttt}\|^2_{2,0} + \nu^{-1}\|p_{tt}\|^2_{2,0} + \|f_{tt}\|^2_{2,0} + \nu\|\nabla u_{tt}\|^2_{2,0} + \nu^{-1}\|\nabla u_{tt}\|^4_{4,0}$$
$$+ \nu^{-1}|||\nabla u|||^4_{4,0} + \nu^{-1}|||\nabla u_{1/2}|||^4_{4,0}).$$

At this point one of the two discrete Gronwall lemmas is applied. The first leads to the time-step restriction that Δt is sufficiently small that $\Delta t < C(\nu^{-3}|||\nabla u|||^4_{\infty,0} + 1)^{-1}$. However, the second may be applied instead and it leads to no restriction on the time-step. This gives

$$\|\phi^h_M\|^2 + \nu\Delta t \sum_{n=0}^{M-1} \|\nabla\phi^h_{n+1/2}\|^2$$
$$\le C\nu^{-1} h^{2k+1}\left(|||u|||^4_{4,k+1} + |||\nabla u|||^4_{4,0}\right)$$
$$+ C\nu^{-1}h^{2k}\left(|||u|||^4_{4,k+1} + \nu^{-1}\|u^h\|^2 + \nu^{-2}|||f|||^2_{2,\star}\right) \tag{9.5.2}$$
$$+ C\nu^{-1}h^{2s+2}|||p_{1/2}|||^2_{2,s+1} + C\nu h^{2k}|||u|||^2_{2,k+1}$$
$$+ C\Delta t^4\left(\|u_{ttt}\|^2_{2,0} + \nu^{-1}\|p_{tt}\|^2_{2,0} + \|f_{tt}\|^2_{2,0}\right.$$
$$+ \nu\|\nabla u_{tt}\|^2_{2,0} + \nu^{-1}\|\nabla u_{tt}\|^4_{4,0}$$
$$\left. + \ \nu^{-1}|||\nabla u|||^4_{4,0} + \nu^{-1}|||\nabla u_{1/2}|||^4_{4,0}\right).$$

The first error estimate in the theorem then follows from the triangle inequality. To obtain the second error estimate in the theorem, we use the last inequality and

$$\begin{aligned}
&\|\nabla\left(u(t_{n+1/2}) - (u^h_{n+1} + u^h_n)/2\right)\|^2 \\
&\leq \|\nabla(u(t_{n+1/2}) - u_{n+1/2})\|^2 + \|\nabla\eta_{n+1/2}\|^2 + \|\nabla\phi^h_{n+1/2}\|^2 \\
&\leq \frac{(\Delta t)^3}{48}\int_{t_n}^{t_{n+1}} \|\nabla u_{tt}\|^2\,dt + Ch^{2k}|u_{n+1}|^2_{k+1} + Ch^{2k}|u_n|^2_{k+1} + \|\nabla\phi^h_{n+1/2}\|^2. \quad \square
\end{aligned}$$

9.6 Remarks on Chapter 9

The next step in reading about solving time-dependent flow problems is the interesting book by Pironneau [74]. Because of the fundamental importance of time-dependent problems there are many excellent CFD books that focus primarily on finite element (and other) methods for the time-dependent NSEs, including the books by Thomasset [96], Quartapelle [78], Ferziger and Peric [33], Orlandi [73], and Baker [8]. A 1976 paper by Baker [9] was one of the first papers on the numerical analysis of the time-dependent NSEs and is still fresh and interesting. It was widely circulated but, sadly, never published. The paper by Baker, Dougalis, and Karakashian [10] was another early, interesting, and fundamental paper. These were soon followed by a series of papers by Heywood and Rannacher [48], [49], [50], [51] which are still considered a large part of the foundation of the area. The extrapolated Crank–Nicholson method has many interesting properties beyond [9], e.g., [5], [89], [47]. The extrapolated trapezoid rule is a good method. It is unconditionally stable and second order accurate and requires the solution of one linear system per time step. Of course, no method is perfect and any time-stepping method that improves upon the extrapolated trapezoid rule is of interest. One such improved time-stepping method which is currently considered to have a good balance between cost, accuracy, and temporal stability is the *fractional step θ-method.* See, for example, [17]. Many practical issues in solving the time-dependent NSE are presented with mathematical and algorithmic clarity by Gresho and Sani [34]. Many more remain unresolved. Reconsidering even the most basic question of stability turns up new and important refinements of methods, e.g., Rebholz [79].

The error analysis of the fully discrete scheme we have given is the classical one. In it, the time discretization is treated as a numerical method for an ODE in time. Consistency is evaluated by Taylor series and stability is proved by summing the difference equations. This is the traditional approach taken in papers on FEM error analysis but it is not clear if it is the "best" one. Various difference methods in time have been recognized as the discrete equations arising from Galerkin discretizations (of various types) of the ODE in time. Once this connection is recognized, many new tools are available for the error analysis, and the error analysis itself is closer to the theory of weak solutions of the NSE. On the other hand, time discretizations are where efficiency begins to become *the* central issue. As a result, the best time discretization for a given problem is often a combination of various basic ones for different terms in the problem. As a result, the field is very complex and requires many diverse tools.

The fact that the (obvious) combination of the stabilization used in Algorithm 9.3 and the extrapolated trapezoid rule (i.e., Algorithm 9.4) is unconditionally stable is developed in [63]. As the Reynolds number increases, laminar flows lose their physical stability and the question of predicting turbulent flows becomes unavoidable and are considered in Chapter 10. In fact, there are many flows that are inhomogeneous combinations of laminar flows in some regions, boundary layer flows, homogeneous turbulence, and anisotropic and inhomogeneous turbulence. Success for these flows will require approaches that are a synthesis of many ideas and adaptivity will certainly be one of the important ingredients, e.g., Hoffman and Johnson [53].

9.7 Exercises

Exercise 102. *The constant $C^*(T, \nu)$ depends exponentially on ν and T. Find this dependence.*

Exercise 103 (error analysis for the Oseen problem). *The Oseen problem is as follows: find $(u(x,t), p(x,t))$ satisfying*

$$u_t + \vec{b} \cdot \nabla u - \nu \Delta u + \nabla p = f, \ \nabla \cdot u = 0 \ in \ \Omega,$$
$$u = 0 \ on \ \partial\Omega, \ u(x,0) = u_0(x) \ in \ \Omega,$$

where the velocity field $\vec{b}$ is smooth, $\nabla \cdot \overrightarrow{b} = 0$, and $\vec{b} = 0$ on $\partial\Omega$. State and prove an error estimate for the semidiscrete method for the Oseen problem.

Exercise 104 (simplifying the error analysis). *Suppose that $||\nabla u|| \in L^\infty(0,T)$, i.e.,*

$$\sup_{0 \le t \le T} ||\nabla u(t)|| < \infty.$$

Rework the error analysis and show how the proof simplifies. How does C^* depend on ν and T in this case?

Exercise 105 (convergence in two dimensions). *Prove convergence in two dimensions without an extra regularity condition on u.*

Exercise 106. *Prove Proposition* 16. ***Hint:*** *First review the corresponding proof in the continuous case.*

Exercise 107 (using the Stokes projection). *Suppose that in the proof of the convergence result the intermediate function $\widetilde{u}$ is chosen to be the Stokes projection of u into V^h:*

$$(\nabla(u - \widetilde{u}), \nabla v^h) = 0 \ \forall v^h \in V^h.$$

Follow the proof and show that the resulting error equation simplifies to

$$(\phi_t^h, v^h) + \nu(\nabla\phi^h, \nabla v^h) = (\eta_t, v^h)$$
$$-(p, \nabla \cdot \phi^h) - [b^*(u,u,v^h) - b^*(u^h,u^h,v^h)].$$

Complete the convergence proof from this point.

Exercise 108 (using the Stokes projection). *Define the Stokes projection $\widetilde{u} \in V^h$ as satisfying*

$$(\nabla(u - \widetilde{u}), \nabla v^h) = 0 \; \forall v^h \in V^h.$$

Show that $\frac{\partial}{\partial t}(\widetilde{u}) = \widetilde{(\frac{\partial u}{\partial t})}$. If v^h in Theorem 22 is chosen to be the Stokes projection, find the resulting rate of convergence of the FEM with the MINI element.

Exercise 109 (dependence on the Reynolds number). *Suppose the flow has only a $O(\mathrm{Re}^{-\frac{1}{2}})$ Prandtl boundary layer near $\partial\Omega$ and is smooth elsewhere. Then, $\nabla u \simeq O(\mathrm{Re}^{+\frac{1}{2}})$ near $\partial\Omega$, in the strip of the width of the boundary layer and is $O(1)$ elsewhere. Similarly, $\nabla\nabla u \simeq O(\mathrm{Re}^{-1})$ in this strip and is smooth elsewhere. Suppose the pressure has no boundary or interior layer but is globally smooth. To estimate the time derivative (very roughly) the NSE state that $u_t + \cdots - \nu\Delta u = \cdots$. Suppose thus that ∇u_t behaves like $\nu\nabla\nabla\nabla u$. Use these to estimate the magnitude of the interpolation error in the right-hand side of the error estimate with respect to the Reynolds number. Draw conclusions!*

Exercise 110 (consistency error of the extrapolated trapezoid rule). *Show that the extrapolated trapezoid rule has second order consistency error.* ***Hint:*** *All that you really need to show beyond the consistency estimate for the usual trapezoid rule is that $E[u(t_n), u(t_{n-1})] = u(t_{n+1/2}) + O(\Delta t^2)$.*

Exercise 111 (unconditional stability). *Prove a stability bound for the extrapolated Crank–Nicholson method without any assumption on the time-step (i.e., unconditional stability).* ***Hint:*** *Set $v^h = u^h_{n+1} + u^h_n$, then mimic the stability bound's proof in the continuous time case.*

Exercise 112 (backward Euler). *Prove unconditional stability of the backward Euler time-stepping method.* ***Hint:*** *Set $v^h = u^h_{n+1}$. Next, adapt this proof to the stabilized, backward Euler method.*

Exercise 113 (stabilized, extrapolated trapezoid rule). (a) *Prove unconditional stability of the stabilized, extrapolated trapezoid rule. (**Hint:** Set $v^h = u^h_{n+1} + u^h_n$, and then look for a new kinetic energy.)* (b) *A practictioner asks, "What is h in Algorithm 9.4?"*

Meshes in use are typically highly anisotropic and vary greatly in size from one region to another. Find an answer to the practictioner's question which both makes some sense and preserves unconditional stability.

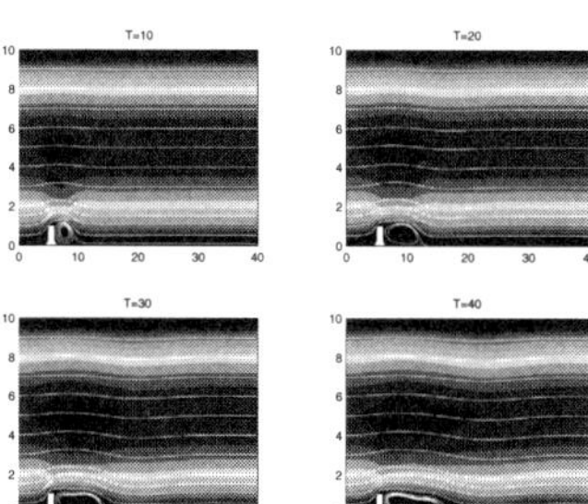

Chapter 10
Models of Turbulent Flow

These considerations justify the view that a considerable mathematical effort towards a detailed understanding of the mechanism of turbulence is called for. The entire experience with the subject indicates that the purely analytical approach is beset with difficulties, which at this moment are still prohibitive. The reason for this is probably as was indicated above: that our intuitive relationship with the subject is still too loose.

J. von Neumann, in *Recent Theories of Turbulence, Collected Works, 1949–1953*, Pergammon Press, New York, 1963.

10.1 Introduction to Turbulence

It remains to call attention to the chief remaining difficulty of our subject.

H. Lamb, in *Hydrodynamics*, Cambridge University Press, Cambridge, UK, 1924.

When I die and go to Heaven there are two matters on which I hope enlightenment. One is quantum electro-dynamics and the other is turbulence. About the former, I am really rather optimistic.

H. Lamb, 1932, quoted in *Computational Fluid Mechanics and Heat Transfer*, by Anderson, Tannehill, and Pletcher, Taylor and Francis, London, 1984.

If a fluid flows over a smooth obstacle and the flow maintains a two dimensional, laminar structure, then the fluid has an $O(\mathrm{Re}^{-1/2})$ transition region (a laminar boundary layer) near the obstacle. For turbulent flows, the structure of transition regions consists of eddies of various sizes. Their structure is more complicated. For example, when air containing streams of smoke to visualize the flow is blown through a screen, each strand of wire sets up a vortex street. After a short distance, these vortices interact and a very complex turbulent flow results. After a further, much greater distance, viscosity drains enough energy from the flow that it relaminarizes.

There are many different conjectures, theories, descriptions, etc., of turbulence. We have already seen one viewpoint: the Leray theory of weak solutions of the NSE. In this theory turbulence is associated with a possible (and yet unproved) breakdown of uniqueness of weak solutions to the NSE. A second theory is that of the turbulent boundary layer along a flat plate or in a pipe. The third theory (which we review in this chapter) is the Kolmogorov theory of homogeneous, isotropic turbulence. This theory is also, so far, unproven mathematically[71] and yet it makes many physical predictions on averages of turbulent flows that are remarkably accurate.

In 1941 Kolmogorov gave this remarkable, universal description of the eddies in turbulent flow by combining a judicious mix of physical insight, conjecture, mathematical analysis, and dimensional analysis, [36], [76]. In his description, the large eddies are deterministic in nature. Those below a critical size are dominated by viscous forces and die very quickly due to these forces. This critical length scale, called the Kolmogorov microscale, is $\eta = O(\text{Re}^{-3/4})$[72] in three dimensions, and $O(\text{Re}^{-1/2})$ for two-dimensional turbulence. To simulate properly the persistent eddies in a three-dimensional flow thus requires taking

$$\Delta x = \Delta y = \Delta z = O(\text{Re}^{-3/4}),$$

giving $O(\text{Re}^{+9/4})$ mesh points in space per time-step. For example, consider the length scales below:

Flow	Re	$l = O(\text{Re}^{-3/4})$	# Mesh points
Model airplane	7×10^4	$10^{-3}\ m$	10^{10}
Subcompact car	6×10^5	$10^{-8}\ m$	10^{12}
Small airplane	2×10^7	$10^{-13}\ m$	10^{16}
Atmospheric flows	Big!!	$1\ mm$	10^{20} and higher.

Such a procedure is called a *direct numerical simulation* (DNS). Clearly, a DNS of many interesting and important flows is not practically possible at the present time. One promising approach to the simulation of turbulent flows is called *large eddy simulation* (LES). The goal of LES is as follows. The users selects a length scale δ of interest. The aim is to simulate accurately the motion of those eddies of size $l \geq O(\delta)$. On the face of it, this seems feasible since the large eddies are believed to be deterministic. The small eddies (accepting Kolmogorov's description) have a universal structure so, in principle, their mean effects on the large eddies should be modelable. The crudest estimate of cost is that we must take

$$\Delta x = \Delta y = \Delta z = O(\delta),$$

with thus $O(\delta^{-3})$ storage required in space per time-step.

There are many other approaches to turbulence and many models in LES other than the one we present in section 10.3. Nevertheless, eddy viscosity models are actually used, with success, in many simulations of turbulent flows.

---

[71] Generally, insofar as something relevant has been proven, the K41 theory has been supported by mathematical analysis.

[72] The length scale of the smallest persistent eddy is traditionally denoted by η rather than l.

10.2 The K41 Theory of Homogeneous, Isotropic Turbulence

The observational material is so large that it allows to foresee rather subtle mathematical results, which would be very interesting to prove.

A. N. Kolmogorov, in Remarks on Statistical Solutions of the Navier-Stokes Equations, *Uspekhi Matematicheskikh Nauk*, 33(1970), 124.

La pensée n'est qu'un au milieu de la nuit. Mais c'est cet éclair qui est tout.

H. Poincaré, in *Science and Reality*, quoted in [21].

Turbulent flow is observed to consist of a cascade of three-dimensional eddies of various sizes. Why do solutions of the NSE exhibit this energy cascade? The answer to this question has been understood since the work of L. F. Richardson and A. N. Kolmogorov and is based on a few fundamental properties of solutions of the NSEs:

- If $\nu = 0$ the total kinetic energy of the flow is exactly conserved,[73]

$$E(u)(t) = E(u)(0) + \int_0^t \frac{1}{L^3} \int_\Omega f \cdot u d\mathbf{x} dt.$$

- The nonlinearity conserves energy globally (since $\int_\Omega u \cdot \nabla u \cdot u dx = 0$) but acts to transfer energy to smaller scales by breaking down eddies into smaller eddies (for example, if $u \simeq (U \sin(\frac{\pi x_1}{l}), 0, 0)^{tr}$ has wave length l and frequency $\frac{\pi}{l}$, then $u \cdot \nabla u \simeq \frac{U^2 \pi}{2l}(\sin(\frac{\pi x_1}{l/2}), 0, 0)^{tr}$ has shorter wave length $\frac{l}{2}$).

- If $\nu > 0$, then the viscous terms dissipate energy from the flow globally:

$$E(u)(t) + \int_0^t \varepsilon(u)(t')dt' = E(u)(0) + \int_0^t \frac{1}{L^3} \int_\Omega f \cdot u d\mathbf{x} dt, \text{ where } \varepsilon(u)(t') \geq 0.$$

- For Re large, the energy dissipation due to the viscous terms is negligible except on very small scales of motion. For example, if $u \simeq (U \sin(\frac{\pi x_1}{l}), 0, 0)^{tr}$, then

$$\text{viscous term on this scale} = -\nu \Delta u \simeq \pi^2 \frac{\nu U}{l^2} \left(\sin\left(\frac{\pi x_1}{l}\right), 0, 0 \right)^{tr}, \text{ from which}$$

$$\text{energy dissipation on this scale} = \varepsilon(u) \simeq \frac{C}{L^3} \frac{\nu U^2}{l^2}.$$

Thus the nonlinear term dominates and the viscous term is negligible if

$$\frac{U^2}{l} >> \frac{\nu U}{l^2}, \text{ i.e., } \frac{lU}{\nu} >> 1.$$

[73] For the physical reasoning in this section it is appropriate to suppose that the energy equality holds and sidestep the deeper questions concerning weak versus strong solutions and energy equality versus energy inequality, e.g., [40], [39].

- The forces driving the flow input energy persistently into the largest scales of motion.

The picture of the energy cascade that results from these effects is thus

Energy is input into the largest scales of the flow.
There is an intermediate range in which nonlinearity drives this energy
into smaller and smaller scales and conserves the global energy because dissipation is negligible.
Eventually, at small enough scales, dissipation is non negligible and
the energy in those smallest scales decays to zero exponentially fast.

This is the physical reasoning behind Richardson's famous description:

> *Big whirls have little whirls*
> *That feed on their velocity,*
> *And little whirls have lesser whirls,*
> *And so on to viscosity.*
> L. F. Richardson, in *Weather Prediction by Numerical Process*, Cambridge University Press, Cambridge, England, 1922.

Inspired by this description, in 1941 Kolmogorov gave a quantitative and universal characterization of the energy cascade (often called the K41 theory). The most important components of the K41 theory are the time (or ensemble) averaged energy dissipation rate, ε, and the distribution of the flows averaged kinetic energy across wave numbers, $E(k)$. Let $\langle \cdot \rangle$ denote long time averaging (e.g., Reynolds [80]),

$$\langle \phi \rangle (\mathbf{x}) := \lim_{T\to\infty} \sup \frac{1}{T} \int_0^T \phi(\mathbf{x}, t) dt.$$

Given the velocity field of a particular flow, $u(\mathbf{x}, t)$, the (time averaged) energy dissipation rate of that flow is thus defined to be[74]

$$\langle \varepsilon \rangle := \lim_{T\to\infty} \sup \int_0^T \frac{1}{L^3} \int_\Omega \nu |\nabla u(\mathbf{x}, t)|^2 d\mathbf{x}\, dt.$$

To present the K41 theory, some preliminaries are necessary to explain how the energy in a velocity field can be decomposed by wave numbers, which is equivalent to the decomposition into length scales.

10.2.1 Fourier series

We consider the NSE under L-periodic boundary conditions.[75] For technical reasons, with periodic boundary conditions, we impose the zero mean condition $\int_\Omega \phi dx = 0$ on $\phi = u, p, f$, and u_0. We can thus expand the fluid velocity in a Fourier series,

$$u(x, t) = \sum_{\mathbf{k}} \widehat{u}(\mathbf{k}, t) e^{-i\mathbf{k}\cdot x}, \text{ where } \mathbf{k} = \frac{2\pi \mathbf{n}}{L} \text{ is the wave number and } \mathbf{n} \in \mathbb{Z}^3.$$

[74] It seems to be unknown if or when this limit exists. In the turbulence community this is often considered an uninteresting mathematical detail since the intuition is that it does exist. Should it not exist, the limit can be replaced by a limit superior (or even a generalized, Banach limit) without changing much of the theory.

[75] The K41 theory does not hold near walls.

It will be useful to modify our vector notation a bit for wave numbers. We will let **k** (bold) denote a wave number vector and k its magnitude, as above. The Fourier coefficients are given by

$$\widehat{u}(\mathbf{k},t) = \frac{1}{L^3}\int_\Omega u(x,t)e^{-i\mathbf{k}\cdot x}dx.$$

Magnitudes of **k**, **n** are defined by

$$|\mathbf{n}| = \{|n_1|^2 + |n_2|^2 + |n_3|\}^{\frac{1}{2}},\ |\mathbf{k}| = \frac{2\pi|\mathbf{n}|}{L},$$
$$|\mathbf{n}|_\infty = \max\{|n_1|, |n_2|, |n_3|\},\ |\mathbf{k}|_\infty = \frac{2\pi|\mathbf{n}|_\infty}{L}.$$

The length scale of the wave number **k** is defined by $l = \frac{2\pi}{|\mathbf{k}|_\infty}$. Parseval's equality implies that the energy in the flow can be decomposed by wave number as follows. For $u \in L^2(\Omega)$,

$$\frac{1}{L^3}\int_\Omega \frac{1}{2}|u(x,t)|^2dx = \sum_{\mathbf{k}} \frac{1}{2}|\widehat{u}(\mathbf{k},t)|^2$$
$$= \sum_k \left(\sum_{|\mathbf{k}|_\infty = k} \frac{1}{2}|\widehat{u}(\mathbf{k},t)|^2\right), \text{ where } \mathbf{k} = \frac{2\pi\mathbf{n}}{L} \text{ is the wave number and } \mathbf{n} \in \mathbb{Z}^3.$$

Recall that $\langle\cdot\rangle$ denotes long time averaging, $\langle\phi\rangle(x) := \limsup_{T\to\infty} \frac{1}{T}\int_0^T \phi(x,t)dt$.

Definition 34. *The kinetic energy distribution functions are defined by*

$$E(k,t) = \frac{L}{2\pi}\sum_{|\mathbf{k}|_\infty = k}\frac{1}{2}|\widehat{u}(\mathbf{k},t)|^2 \text{ and}$$
$$E(k) := \langle E(k,t)\rangle.$$

Parseval's equality thus can be rewritten as

$$\frac{1}{L^3}\int_\Omega \frac{1}{2}|u(x,t)|^2dx = \frac{2\pi}{L}\sum_k E(k,t) \text{ and}$$
$$\left\langle \frac{1}{L^3}\int_\Omega \frac{1}{2}|u(x,t)|^2dx\right\rangle = \frac{2\pi}{L}\sum_k E(k).$$

10.2.2 The inertial range

For the locally isotropic turbulence the [velocity fluctuation] distributions F_n are uniquely determined by the quantities ν, the kinematic viscosity, and ε, the rate of average dispersion of energy per unit mass [energy flux]....

For pulsations [velocity fluctuations] of intermediate orders where the length scale is large compared to the scale of the finest pulsations, whose energy *is directly dispersed into heat due to viscosity, the distribution laws F_N are uniquely determined by F and do not depend on ν.*

I. Kolmogorov, statement of the first and second hypothesis of similarity in The local structure of turbulence in incompressible viscous fluids for very large Reynolds number, *Doklady Akademii Nauk SSSR* 30 (1941), 9–13.

The K41 theory states that at high enough Reynolds numbers there is a range of wave numbers,

$$0 < k_{\text{min}} := U\nu^{-1} \leq k \leq \langle \varepsilon \rangle^{\frac{1}{4}} \nu^{-\frac{3}{4}} =: k_{\text{max}} < \infty, \tag{10.2.1}$$

known as the inertial range, beyond which the kinetic energy in a turbulent flow is negligible, and in this range

$$E(k) \doteq \alpha \langle \varepsilon \rangle^{\frac{2}{3}} k^{-\frac{5}{3}}, \tag{10.2.2}$$

where α is the universal Kolmogorov constant whose value is generally believed to be between 1.4 and 1.7, k is the wave number, and ε is the particular flow's energy dissipation rate. In this formula, the energy dissipation rate $\langle \varepsilon \rangle$ is the only parameter which differs from one flow to another. Indeed, in Pope [76], Figure 6.14 the power spectrums of 17 different turbulent flows are plotted on log-log plots. The slope of the linear region in this plot has the universal value of $-\frac{5}{3}$ for all 17 turbulent flows, exactly corresponding to the $k^{-\frac{5}{3}}$ law.

We present now this argument of Kolmogorov. It begins with a physical conjecture.

Conjecture 1. *The time-averaged kinetic energy depends only on the time averaged energy dissipation rate ε and the wave number k.*

Beginning with this, postulate a simple power law dependency of the form

$$E(k) \simeq C \langle \varepsilon \rangle^a k^b. \tag{10.2.3}$$

If this relation is to hold the units, denoted by $[\cdot]$ on the left-handed side must be the same as the units on the right-handed side, $[LHS] = [RHS]$. The three quantities in the above have the units[76]

$$[k] = \frac{1}{\text{length}}, \; [\langle \varepsilon \rangle] = \frac{\text{length}^2}{\text{time}^3}, \; [E(k)] = \frac{\text{length}^3}{\text{time}^2}.$$

Inserting these units into the above relation gives

$$\frac{\text{length}^3}{\text{time}^2} = \frac{\text{length}^{2a}}{\text{time}^{3a}} \frac{1}{\text{length}^b} = \text{length}^{2a-b}\text{time}^{-3a}, \text{ giving}$$

$$3a = 2, \, 2a - b = 3, \text{ or } a = \frac{2}{3}, \, b = -\frac{5}{3}.$$

Thus, Kolmogorov's law follows:

$$E(k) = \alpha \langle \varepsilon \rangle^{\frac{2}{3}} k^{-\frac{5}{3}} \text{ over the inertial range } 0 < k \leq C(L \,\text{Re}^{-\frac{3}{4}})^{-1}.$$

[76]Note that from the definitions of E and $E(k)$ we have $[E] = \frac{\text{length}^2}{\text{time}^2}$ and $[E(k)] = \text{length} \times [E] = \frac{\text{length}^3}{\text{time}^2}$.

The above estimate $\eta \sim L\,\mathrm{Re}^{-\frac{3}{4}}$ for the Kolmogorov microscale is derived by similar physical reasoning. Let the reference large scale velocity and length (which are used in the definition of the Reynolds number) be denoted by U, L. At the scales of the smallest persistent eddies (the bottom of the inertial range) we shall denote the smallest scales of velocity and length by v_{small}, η. We form two Reynolds numbers:

$$\mathrm{Re} = \frac{UL}{\nu},\ \mathrm{Re}_{small} = \frac{v_{small}\eta}{\nu}.$$

The global Reynolds number measures the relative size of viscosity on the large scales and when Re is large the effects of viscosity on the large scales are then negligible. The smallest scales Reynolds number similarly measures the relative size of viscosity on the smallest persistent scales. Since it is nonnegligible we must have

$$\mathrm{Re}_{small} \simeq 1, \text{ equivalently } \frac{v_{small}\eta}{\nu} \simeq 1.$$

Next comes an assumption of statistical equilibrium:

energy input at large scales = energy dissipation at smallest scales.

The largest eddies have energy which scales like $O(U^2)$ and associated time scale $\tau = O(\frac{L}{U})$. The rate of energy transfer/energy input is thus $O(\frac{U^2}{\tau}) = O(\frac{U^3}{L})$.[77] The small scales energy dissipation from the viscous terms scales like

$$\varepsilon_{small} \simeq \nu|\nabla u_{small}|^2 \simeq \nu\left(\frac{v_{small}}{\eta}\right)^2.$$

Thus we have the second ingredient:

$$\frac{U^3}{L} \simeq \nu\left(\frac{v_{small}}{\eta}\right)^2.$$

Solving the first equation for v_{small} gives $v_{small} \simeq \frac{\nu}{\eta}$. Inserting this value for the small scales velocity into the second equation, solving for the length scale η, and rearranging the result in terms of the global Reynolds number gives the following estimate for η which determines the above estimate for the highest wave number in the inertial range:

$$\eta = \eta_{Kolmogorov} \simeq \mathrm{Re}^{-\frac{3}{4}}L.$$

This estimate for the size of the smallest persistent solution scales is the basis for the estimates of $O(\mathrm{Re}^{\frac{9}{4}})$ mesh-points in space leading to complexity estimates of $O(\mathrm{Re}^3)$ for DNS of turbulent flows.

[77] It is known for many turbulent flows that as predicted by K41, ε scales like $\frac{U^3}{L}$. This estimate expresses statistical equilibrium in K41 formalism [36], [76].

10.3 Models in Large Eddy Simulation

The distinguishing feature of a turbulent flow is that its velocity field appears to be random and varies unpredictably. The flow does, however, satisfy the Navier–Stokes differential equations, which are not random. This contrast is the source of much of what is interesting in turbulence theory....

One should keep in mind that a practical person is usually interested only in mean properties of a small number of functionals of the flow (e.g., lift and drag in the case of flow past a wing), and these could conceivably be obtained even when the details of the flow are unknown.

A. J. Chorin, in *Lectures on Turbulence Theory*, Publish or Perish, Houston, 1975.

The upper limit to the size of an eddy is, like the length of a piece of string, a matter of human convenience.

L. F. Richardson, in [81].

Eddy-viscosity LES modeling is motivated by two physical ideas:

1. *Differing dynamics of the large and small eddies.* At high Reynolds number the fluid velocity is exponentially sensitive to perturbations of the problem data. This sensitivity, however, is not uniform. The large structures (large eddies) evolve deterministically and are thus not sensitive [11]. The small eddies are sensitive because they have a random character. Their random character does, however, have universal features so that there is hope that their mean effects on the large eddies can be modelled.

2. *The Eddy-Viscosity Hypothesis/Boussinesq Assumption.* The eddy viscosity hypothesis/ Boussinesq assumption is that

The small eddies act to drain energy from the large eddies.

This description is quite old in fluid mechanics. For example, Venturi in 1797 wrote about retardation of a flow caused by eddies of different velocities interacting,[78] and Boussinesq gave an early and compelling theoretical argument in favor of it. Paraphrasing Prandtl's 1931 description, the eddy viscosity hypothesis is that a

> fluid flowing turbulently behaves, in the mean, like a fluid of greatly increased and highly variable viscosity.

The *eddy viscosity hypothesis* thus seeks to model the energy lost to the resolved scales when two eddies interact and break down into ones smaller than the cutoff length scale. This process of breakdown of eddies into smaller and smaller ones is often called the Richardson energy cascade since it was described by Richardson in his famous poem. There is considerable evidence in real fluids for those two physical ideas. Their mathematical derivation from the NSE is still, however, an open question. We shall use these two ideas to present the family of models we are considering. Let $g(x)$ be a smooth function with

[78] In 1838 Saint-Venet called this extra retardation *extraordinary friction*, and others had called it *loss of live force*, where *live force* was the term used for kinetic energy.

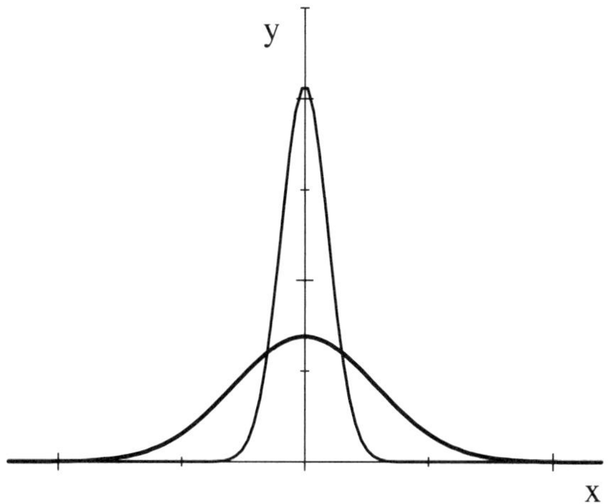

Figure 10.1. *A Gaussian filter (heavy) and rescaled (thin).*

$0 \leq g \leq 1$, $g(0) = 1$, and $\int_{\mathbb{R}^d} g\, dx = 1$. The mollifier $g_\delta(x)$ is defined (as usual, e.g., [14]) by

$$g_\delta(x) := \delta^{-d}\, g(x/\delta).$$

The function $g(x)$ is assumed to have compact support or, at least, fast (e.g., exponential) decay. A typical example is a Gaussian,

$$g(x) = (6/\pi)^{d/2} \exp(-6\, x_j\, x_j),$$

whose plot is given in Figure 10.1 for $\delta = 1$ (bold curve) and rescaled by $\delta = \frac{1}{3}$ (thin curve).

The local spacial filter is defined by convolution with g_δ. Thus, given $u(x)$ define

$$\bar{u}(x) = (g_\delta * u)(x) := \int_{\mathbb{R}^d} g_\delta(x - y)u(y)dy \text{ and } u' = u - \bar{u},$$

where u is extended by zero outside of Ω. This convolution is a smoothing operator. It eliminates any local oscillations smaller than $O(\delta)$. Convolution has many other important mathematical properties. For example, it commutes with differentiation so

$$g_\delta * \left(\frac{\partial}{\partial x_j} u\right) = \frac{\partial}{\partial x_j}(g_\delta * u),$$

and it is smoothing if the filter is itself smooth. A smoothing averaging process is often depicted (Figure 10.2) as decomposing a wiggly function, u, into its mean (bold curve below), $\overline{u}$, and its fluctuation about the mean (the dashed curve fluctuating about the x-axis), $u' := u - \overline{u}$.

Exact spacial averaging suppresses small spacial scales. It is widely believed that spacial averaging (through the space and time coupling in the equations of motion) also suppresses correspondingly small time scales. This belief certainly corresponds to our everyday observation of fluids in motion: spacial averaging (done by looking at a river from a greater distance) seems to make it flow more slowly. This was well described by Wordsworth:

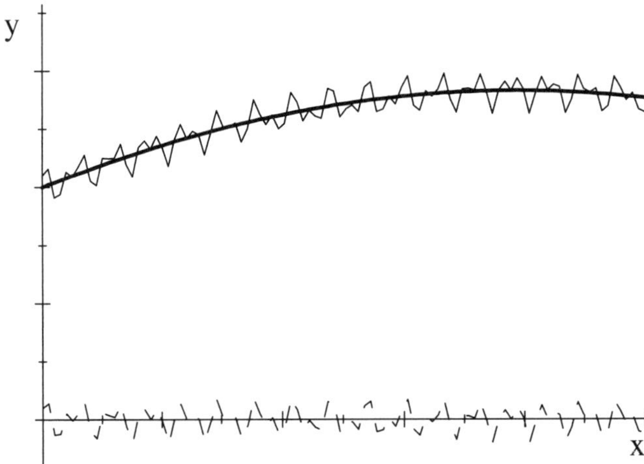

Figure 10.2. *A curve, its mean (heavy line) and fluctuation (dashed).*

Yon foaming flood seems motionless as ice,
Its dizzy turbulence eludes the eye,
Frozen by distance.
W. Wordsworth, 1770–1850, in "Address to Kilchurn Castle."

Ignoring boundaries for the moment and using these nice properties of convolution, apply this filter to the NSE (i.e., take $g_\delta * \text{ NSE}(u) = g_\delta * f$). This gives the SFNSE = space-filtered-NSE,

$$\bar{u}_t + \nabla \cdot (\bar{u}\,\bar{u}) + \nabla \bar{p} - \text{Re}^{-1}\Delta \bar{u} + \nabla \cdot \mathbf{R}(u, u) = \bar{f}, \ \nabla \cdot \bar{u} = 0, \tag{10.3.1}$$

where $\mathbf{R}(u, u)$ is the tensor representing the stress the unresolved scales exerts upon the resolved scales:

$$\mathbf{R}(u, u) := \overline{u\,u} - \bar{u}\,\bar{u}.$$

The tensor $\mathbf{R}(u, u)$ is often called the *subfilter scale stress tensor* and it is sometimes called the Reynolds stress tensor. (There is a terminological debate about whether the latter is correct or a misnomer. It is also called the *subfilter scale stress tensor*, a much more accurate term.) Since $\mathbf{R}$ is a function of u and not only of $\bar{u}$, this system is not closed. The simplest closure assumption is the eddy viscosity hypotheses (given above) that the turbulent stress is a linear function of the large scales' deformation tensor $\mathbf{D}(\bar{u})$. This model is

$$\nabla \cdot \mathbf{R}(u, u) \sim -\nabla \cdot (2\nu_T\, \mathbf{D}(\bar{u})), \tag{10.3.2}$$
$$\nu_T \ := \ \text{turbulent viscosity coefficient.}$$

In fact, the substitution (10.3.2) is not strictly correct. Just as for the usual stress tensor, $\mathbf{R}$ must be split into two parts. One, the average of the stresses in the $x - y - z$

directions, is incorporated into the turbulent pressure and the other, the trace free part of **R**, is modeled by turbulent diffusion,

$$\mathbf{R} = \left(\mathbf{R} - \frac{1}{3}\text{trace}(\mathbf{R})\mathbf{I}\right) + \frac{1}{3}\text{trace}(\mathbf{R})\mathbf{I}.$$

Thus the mean turbulent pressure $\bar{p}$ in (10.3.1) should really be adjusted by

$$\bar{p} \Leftarrow \bar{p} + \frac{1}{3}\text{trace}(\mathbf{R}),$$

and eddy viscosity really models only the zero trace part of **R**,

$$\nabla \cdot \left(R - \frac{1}{3}\text{trace}(R)I\right) \sim -\nabla \cdot (\nu_T D(\bar{u})).$$

There are many models for determining the turbulent viscosity coefficient,

$$\nu_T = \nu_T(\delta,\ \bar{u}).$$

Recall that (10.3.2) is a modeling assumption. Thus, when (10.3.2) is used the solution sought is no longer $(\bar{u},\ \bar{p})$ but rather an approximation to $(\bar{u},\ \bar{p})$. Calling that approximation (w, q), the model that results is

$$w_t + \nabla \cdot (ww) + \nabla q - \nabla \cdot (2(\text{Re}^{-1} + \nu_T)\mathbf{D}(w)) = \bar{f} \text{ in } \Omega, \qquad (10.3.3)$$
$$\nabla \cdot w = 0 \text{ in } \Omega.$$

10.3.1 A first choice of ν_T

With four parameters I can fit an elephant, and with five I can make him wiggle his trunk.

J. von Neumann, quoted by Freeman Dyson in "A Meeting with Enrico Fermi", *Nature* 427 (2004), 297.

The simplest choice of ν_T is a constant. This is not a good choice because it just reduces a turbulent flow to a laminar one. However, it introduces some interesting ideas, so we shall consider it first. One goal of an LES model is to make the eddies smaller than $O(\delta)$ "vanish." With this in mind, the model (10.3.3) describes a flow with effective Reynolds number

$$\text{effective Reynolds number} := (\text{Re}^{-1} + \nu_T)^{-1} \sim \nu_T^{-1}.$$

The K41 theory of Kolmogorov thus predicts the length scale of the smallest persistent eddy in solutions of the model to be $\eta_{LES} \doteq O([\text{Re}^{-1} + \nu_T]^{3/4})$. We wish $\eta_{LES} = O(\delta)$, so, in the interesting case in which $\text{Re}^{-1} << O(\delta)$, this means $\nu_T = O(\delta^{4/3})$. This yields the global choice $\nu_T = C\delta^{4/3}$, where $C > 0$ is an $O(1)$ fitting constant.

The problem with this constant choice of ν_T is the model's action on the smooth parts of the flow—it is just too diffusive! The flow that results is laminar. This observation motivates efforts to find better models which have similar effects on the small eddies but do not distort the large eddies too much.

10.4 The Smagorinsky Model for ν_T

The road to full knowledge of the variations of viscosity appears to lie in the study of diffusion of eddies.
L. F. Richardson, 1922, in [81].

The only encouraging prospect is that current progress in understanding turbulence will restrict the freedom of such modeling and guide these efforts toward a more reliable discipline.
H. W. Liepman, in *American Scientist*, 62 (1979), p. 221.

In 1963 Smagorinsky proposed the choice for ν_T which is still the most popular today:

$$\nu_T = (C_S\, \delta)^2 |\mathbf{D}(w)|. \tag{10.4.1}$$

This model was advanced independently by Ladyzhenskaya for other reasons. The same regularization of the compressible NSE had been used in the 1950s by Richtmeyer and von Neumann for computations of compressible flows with shocks. It is also the most mathematically appealing choice for ν_T.

Let us now consider a choice like the above,

$$\nu_T = C_s\, \delta^r\, |\mathbf{D}(w)|^s,$$

where r and s must be determined. Following the previous reasoning, the *effective local Reynolds number* is

$$\mathrm{Re}_{Effective} = (\mathrm{Re}^{-1} + C_s\, \delta^r |\mathbf{D}(w)(x)|^s)^{-1} \sim (C_s\, \delta^r |\mathbf{D}(w)(x)|^s)^{-1}.$$

If we assume the K41 theory can be applied *locally* as well as globally after time averaging (and there is little evidence either for or against this assumption, but to make progress something must be assumed), this gives the *local* length scale of

$$\eta_{LES}(x) \doteq O([\mathrm{Re}^{-1} + C_s\, \delta^r |\mathbf{D}(w)(x)|^s]^{3/4}) \doteq O(\delta^{(3/4)r} |\mathbf{D}(w)|^{3/4\, s}(x)),$$

in the most interesting case when Re^{-1} is negligible.

If x is chosen to lie in an eddy of size $\leq O(\delta)$, then $|\mathbf{D}(w)(x)| \geq O(\delta^{-1})$. Thus, if we want

$$\min_x \eta_{LES}(x) = \delta,$$

this gives

$$\delta \doteq \delta^{3/4r}\, \delta^{-3/4s} \text{ or } r = \frac{4}{3} + s.$$

This choice of r and s gives an attractive family of LES models,

$$\nu_T(\delta, w) := C_s\, \delta^{4/3+s} |\mathbf{D}(w)|^s, \text{ where } s > 0, \tag{10.4.2}$$

yielding the model

$$\begin{aligned} &w_t + \nabla \cdot (w\, w) + \nabla q - \nabla \cdot [2(\mathrm{Re}^{-1} + C_s\, \delta^{4/3+s} |\mathbf{D}(w)|^s)\, \mathbf{D}(w)] = \bar{f} \text{ in } \Omega, \\ &\nabla \cdot w = 0, \text{ in } \Omega. \end{aligned} \tag{10.4.3}$$

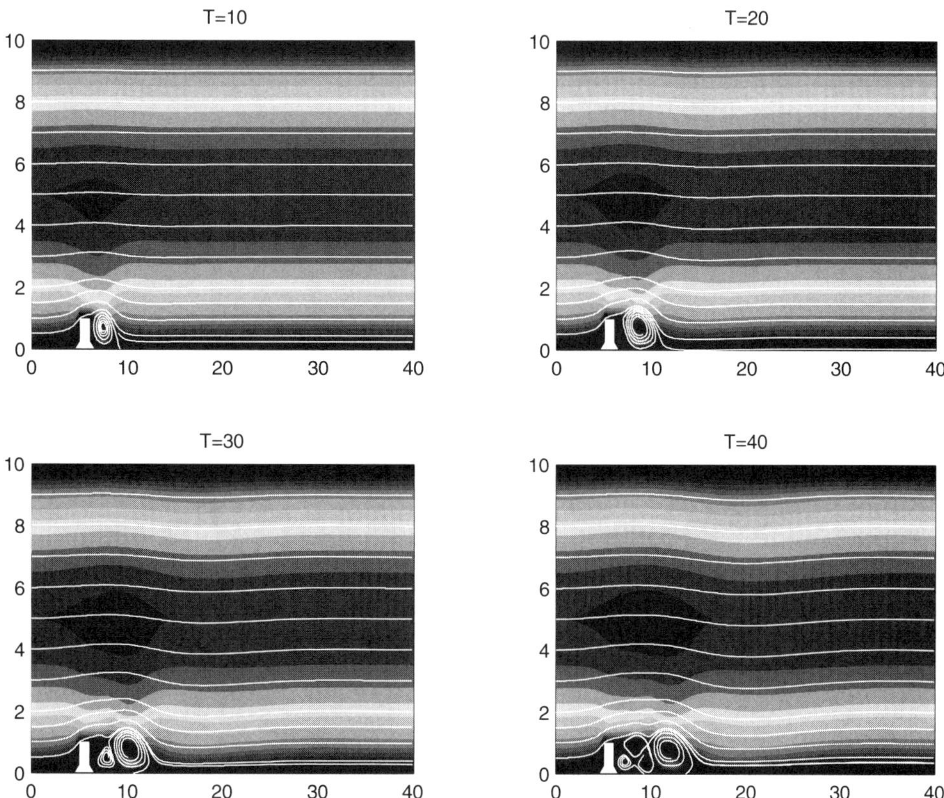

Figure 10.3. *Eddies are shed and roll down channel.*

For the large eddies (when $|\mathbf{D}(w)| = O(1)$), we can make the influence of the turbulent viscosity smaller by picking s larger. We expect that the eddies smaller than $O(\delta)$ are also rapidly attenuated. Thus, (10.4.3) is an important improvement in LES models over constant ν_T.

The Smagorinsky model corresponds to special choice $r = 2, s = 1$. It is thus given by

$$\begin{aligned} &w_t + \nabla \cdot (w\, w) + \nabla q - \nabla \cdot [2(\, Re^{-1} + (C_S\, \delta)^2 |\mathbf{D}(w)|)\, \mathbf{D}(w)] = \bar{f}, \\ &\nabla \cdot w = 0 \text{ in } \Omega. \end{aligned} \tag{10.4.4}$$

Variations on the Smagorinsky model have proved to be the workhorse in large eddy simulations of industrial flows. Variants are needed because the model as presented, while far better than constant eddy viscosity, is still far too dissipative. For example, here is a simple test of the Smagorinsky model for two-dimensional flow over a step (far from the case of turbulence). At the Reynolds number of this test, around 600, eddies are periodically shed from the step and roll down the channel (Figure 10.3); the flow does not approach a steady state.

However, if this simulation is performed on a coarser mesh, it makes sense to try the Smagorinsky model to see if it gives the correct large structures. Here is the result which is

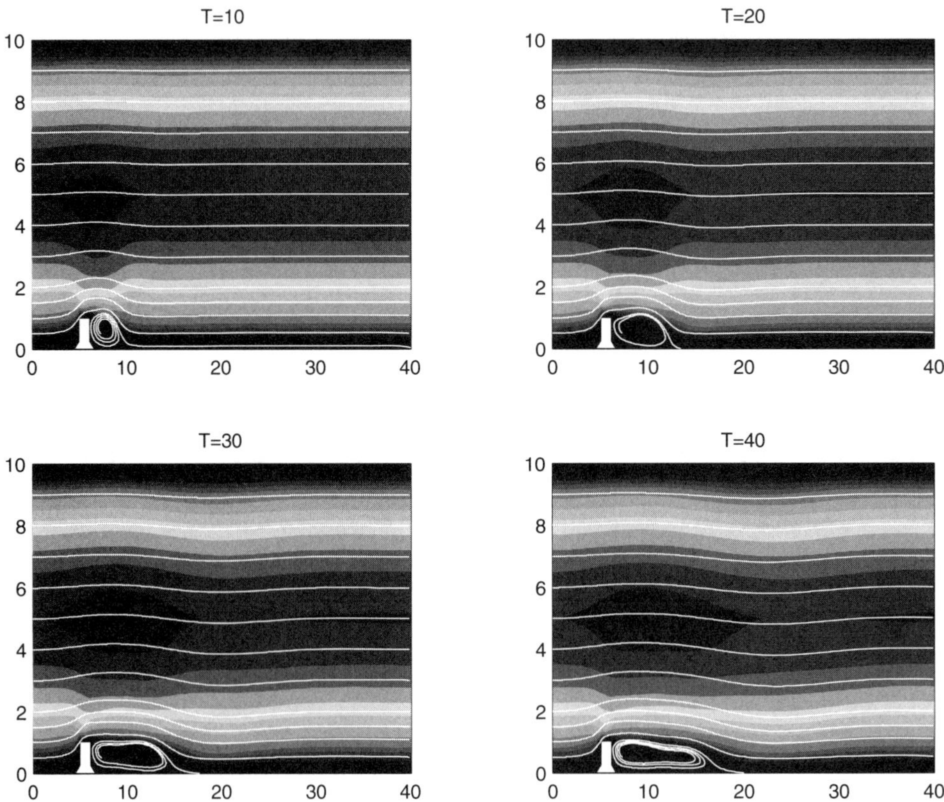

Figure 10.4. *Smagorinsky model predicts flow reaches equilibrium quickly.*

clearly overdiffused: eddies are not shed and the model predicts the flow will quickly reach a nonphysical equilibrium with one long, attached eddy behind the step (Figure 10.4).

Some modifications used to reduce the effects of eddy viscosity on the dynamics of the large structures are the following:

1. taking the Smagorinsky constant small, typically $C_S \approx O(0.1)$;
2. Van Driest damping: taking $C_S = C_S(x)$, where $C_S(x) \to 0$ rapidly as $x \to \partial\Omega$;
3. the dynamic model of Germano: fitting $C_S(x)$ adaptively to other relations and thus even allowing for negative viscosities.

10.5 Near Wall Models: Boundary Conditions for the Large Eddies

It is disappointing to find that the boundary regions in large-eddy simulations contain serious errors. This cannot, however, be considered too surprising, as close to the surface the potential rationality of the large-eddy simulation

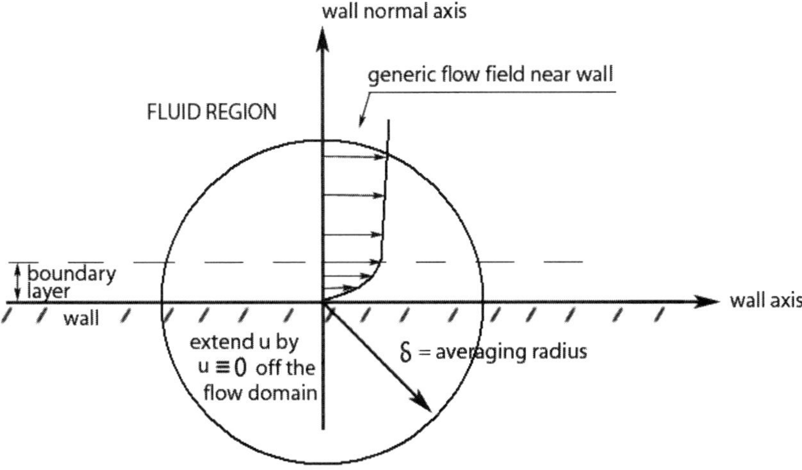

Figure 10.5. *$\bar{u}$ does not vanish on $\partial\Omega$.*

vanishes as the dominant eddy-scales become comparable with, and smaller than, the filter-scale.

P. J. Mason, in *Large Eddy Simulation: A Critical Review of the Technique*, Quarterly Journal Royal Meteorological Society 120 (1994), 1–26.

Typical flow geometries in turbulent flow have inflow boundaries, obstacles, walls, and outflow boundaries, and correct boundary conditions must be imposed for all of them. Even the simplest case (which we are considering) of an internal flow problem is already not so simple, as we shall see! Let $\bar{u} = g_\delta * u$, $\bar{p} = g_\delta * p$, and let (w, q) approximate $(\bar{u}, \bar{p})$ and satisfy the initial value problem:

$$\begin{aligned} &w_t + \nabla \cdot (w\, w) + \nabla q - \mathrm{Re}^{-1}\Delta w - \nabla \cdot (\nu_T \mathbf{D}(w)) = \bar{f} \text{ in } \Omega \times (0, T], \\ &\nabla \cdot w = 0 \text{ in } \Omega \times (0, T], \text{ and} \\ &w(x, 0) = \bar{u}_0(x), \text{ in } \Omega. \end{aligned}$$

In practical calculations, the most commonly used boundary condition for w is simply

$$w = 0 \text{ on } \partial\Omega. \tag{10.5.1}$$

Recall that $w \cong \bar{u}$. Consider Figure 10.5; it is clear that imposing the no-slip condition (10.5.1) on the large eddies introduces a consistency problem since $\bar{u} = g_\delta * u \neq 0$ on $\partial\Omega$: $u = 0$ on $\partial\Omega$ but $\bar{u}$ being an $O(\delta)$ average of u, does *not* vanish on $\partial\Omega$.

It is also clear that large eddies do not stick to boundaries if you consider common examples like tornadoes and hurricanes. When contacting the earth, these do move along the earth and lose energy as they move. The picture painted by boundary layer theory and these last examples give important clues to finding the correct boundary conditions to be *no-penetration* and *slip with friction*. Mathematically, these are written as

$$\begin{cases} w \cdot \hat{n} = 0 \text{ on } \partial\Omega \text{ and} \\ w \cdot \hat{\tau}_j + \beta(\delta, \mathrm{Re})\hat{n} \cdot \Pi(w) \cdot \hat{\tau}_j = 0 \text{ on } \partial\Omega, \end{cases} \tag{10.5.2}$$

where $(\hat{n}, \hat{\tau}_j)$ are the unit tangent and normal vectors and $\Pi(w)$ is the stress tensor associated with the viscous and turbulent stresses:

$$\hat{n} \cdot \Pi \cdot \hat{\tau}_j = \hat{n} \cdot (2\,\mathrm{Re}^{-1}\mathbf{D}(w) + \nu_T \mathbf{D}(w)) \cdot \hat{\tau}_j.$$

The modeling problem now reduces to identification of the (linear or even nonlinear) friction coefficient β. The analysis of Maxwell (see the discussion in Chapter 5) gives some insight into the friction coefficient. If we identify the microlength scale with the filter length scale δ, then this suggests

$$\beta \sim \mathrm{Re}\frac{\delta}{L}.$$

With this scaling, as $\delta \to 0$ we recover the no-slip condition and as $\mathrm{Re} \to \infty$, we transition to free-slip and the Euler equations. Thus, this passes the test of asymptotic reasonableness. Of course, we do not compute in either asymptotic regime, so a good formula must be found for the friction coefficient in the intermediate region. For initial attempts, see [42] and [60].

10.6 Remarks on Chapter 10

Fortunately, there are several excellent books connecting mathematics with the basic physical understanding of turbulence, including the books by Frisch [36] and Pope [76]. There are many experimental studies of turbulence pointing to the (still unproved, although progress continues to be made [35], but widely believed to be true) idea that the large scales evolve deterministically while sensitive dependence is restricted to the small turbulent fluctuations; see, for example, [11]. The book by Mohammadi and Pironneau [72] gives an excellent presentation of the numerical analysis of the k-epsilon model. For good books on large eddy simulation, see [84], [57], and [14]. Numerical analysis of the Smagorinsky model was completed in the paper by John and Layton. It is related to work on the Ladyzhenskaya model of Du and Gunzburger [27]. There are many models in LES that show great promise; see, e.g., [14], [84]. Most are derived to improve various features of the Smagorinsky model and compared with it in computational tests. At this time, approximate deconvolution models appear to be among the most promising. Briefly, the SFNSE (10.3.1) can be rearranged to read

$$\bar{u}_t + \nabla \cdot (\overline{uu}) + \nabla \bar{p} - \mathrm{Re}^{-1}\Delta \bar{u} = \bar{f},\ \nabla \cdot \bar{u} = 0.$$

The filtering operation $u \to \overline{u}$ is not stably invertible. However, there are many ways to calculate approximate inverses. Let the operator D denote a (bounded) approximate inverse obtained from one of the methods for solving ill-posed problems. Then, $u \simeq D(\overline{u})$, and the SFNSE can be written

$$\bar{u}_t + \nabla \cdot (\overline{D(\bar{u})D(\bar{u})}) + \nabla \bar{p} - Re^{-1}\Delta \bar{u} = \bar{f}$$
$$+ \text{ the error in approximate deconvolution.}$$

If the deconvolution error is small, dropping it results in a closed model for the flow averages. As an example, recall from Chapter 9 that linear extrapolation in time gives $u(t) = 2u(t - \Delta t) - u(t - 2\Delta t) + O(\Delta t^2)$. One can view $\overline{u}$ as a function of δ (this is scale space) and extrapolate linearly down to $\delta = 0$,i.e., $u = \overline{u}(\delta)|_{\delta=0}$. Linear extrapolation in scale space gives $u = 2\overline{u} - \overline{\overline{u}} + O(\delta^4)$. Thus, take $D(\overline{u}) = 2\overline{u} - \overline{\overline{u}}$ as one example. Currently there are some very interesting ideas in the area that come from viewing modeling and discretization together. These approaches include the variational multiscale method (an idea of T. Hughes and his coworkers, e.g., [55]) and a synthesis of turbulence modeling, numerical regularization, and adaptivity, e.g., Hoffman [52] and Hoffman and Johnson [53].

There are also many attempts to finding effective boundary conditions for the large eddies. The one described herein is from [42] and [60]. Finding good near wall models that provide accuracy in the mean flow without resolving the boundary layers is a very important and difficult problem in CFD.

10.7 Exercises

Exercise 114 (dimensional analysis of inequalities in analysis). *A rough idea of the dependence of various constants in analysis can be obtained by considering the dimensions of the terms involved. For example, taking the dimensions of both sides of the Poincaré–Friedrichs inequality gives*

$$\int_\Omega |u|^2 dx \lesssim C_{PF}^2 \int_\Omega |\nabla u|^2 dx \textit{ suggests}$$

$$\left[\int_\Omega |u|^2 dx\right] \approx [C_{PF}]^2 \left[\int_\Omega |\nabla u|^2 dx\right].$$

To simplify, let U denote units of velocity. We then have

$$L^3 U^2 \approx [C_{PF}]^2 L^3 \left(\frac{U}{L}\right)^2,$$

so we expect C_{PF} to scale like $L = \operatorname{diam}(\Omega)$. Pick five inequalities in analysis which you feel are key to the numerical analysis of the NSE. Repeat the above heuristic analysis to get insight into the scaling of their constants.

Exercise 115 (hyperviscosity). *Consider the NSE under periodic boundary conditions with the viscous term replaced by a fourth order, hyperviscosity term:*

$$\begin{aligned} u_t + u \cdot \nabla u + \nabla p + \nu \Delta^2 u &= f \textit{ in } \Omega \times (0, T), \\ u(x, 0) &= u_0(x) \textit{ in } \Omega, \textit{ and} \\ \nabla \cdot u &= 0 \textit{ in } \Omega \times (0, T). \end{aligned}$$

Show that the solution of the above satisfies an energy equality and argue from it that its solution has an energy cascade. Assuming such a cascade, find the microscale and predict the energy spectrum. Hyperviscosity models are commonly used in geophysical flow simulations.

Exercise 116 (a modified kinetic energy). *Consider the NSE under periodic boundary conditions and with a modification of the kinetic energy:*

$$\begin{aligned} u_t - \delta^2 \Delta u_t + u \cdot \nabla u + \nabla p - \nu \Delta u &= f \text{ in } \Omega \times (0, T), \\ u(x, 0) &= u_0(x) \text{ in } \Omega, \text{ and} \\ \nabla \cdot u &= 0 \text{ in } \Omega \times (0, T). \end{aligned}$$

What is the connection between this fluids model and the stabilized backward Euler timestepping method of Chapter 9? Show that the solution of the above satisfies an energy equality and argue from it that its solution has an energy cascade. Assuming such a cascade, find the microscale and predict the energy spectrum.

Exercise 117 (differential filters). *Another filter is the differential filter proposed by Germano* [43]. *Briefly, given an L-periodic $\phi(x)$, its average $\overline{\phi}$ is the unique L-periodic solution of*

$$-\delta^2 \Delta \overline{\phi} + \overline{\phi} = \phi \text{ in } \Omega.$$

Show that the differential filter is within $O(\delta^2)$ of the Gaussian filter. Average the NSE with this differential filter and derive the differentially SFNSE.

Exercise 118 (reversibility versus irreversibility). *Eddy viscosity is a description of the global effects of the nonlinear interactions in fluid flows. As such, it has many quantitative (predictive) flaws and some descriptive flaws as well. For example, verify that the true Reynolds/subfilter scale stresses, $\mathbf{R}(u, u) := \overline{u\,u} - \bar{u}\,\bar{u}$, satisfy*

$$\mathbf{R}(-u, -u) = \mathbf{R}(u, u).$$

(This is known as reversibility.) On the other hand, show that the eddy viscosity approximation, $\mathbf{R}(u, u) \sim -2\nu_T\, \mathbf{D}(\bar{u})$ (+ pressure term), is irreversible:

$$2\nu_T\, \mathbf{D}(-\bar{u}) = -2\nu_T\, \mathbf{D}(\bar{u}).$$

Exercise 119. *The theory of two-dimensional turbulence predicts that $\eta = O(\mathrm{Re}^{-\frac{1}{2}})$. Show that the analogous choice is $\nu_T = C\delta^2$ for two-dimensional turbulence.*

Exercise 120. *Consider two-dimensional turbulence. Repeat the analysis and find the width of transition regions for the Smagorinsky model. Are transition regions in solutions spread to $O(\delta)$ in two dimensions?*

Exercise 121. *Find the units of C_S in the Smagorinsky model.*

Exercise 122. *Find the variational formulation of the Smagorinsky model under no-slip boundary conditions: $w = 0$ on $\Gamma = \partial\Omega$ and under the slip with friction boundary condition* (10.5.2).

Appendix

Nomenclature

A.1 Vectors and Tensors

Vector dot product: $u \cdot v := u_1 v_1 + u_2 v_2 + u_3 v_3 (= u_i v_i)$
Tensor contraction: $\mathbf{T} : \mathbf{S} := \sum_{i=1}^{3} \sum_{j=1}^{3} \mathbf{T}_{ij} \mathbf{S}_{ij} (= \mathbf{T}_{ij} \mathbf{S}_{ij})$
Gradient tensor: $(\nabla u)_{ij} = \frac{\partial u_j}{\partial x_i} = u_{j,i}$
Symmetric part of the gradient is the deformation tensor:

$$\mathbf{D}(u)_{ij} := \frac{1}{2}(\nabla u + \nabla u^t) = \frac{1}{2}(u_{j,i} + u_{i,j}) = \nabla^s u$$

Skew symmetric parts of the gradient is the spin tensor:

$$\Omega(u)_{ij} := \frac{1}{2}(\nabla u - \nabla u^t) = \frac{1}{2}(u_{j,i} - u_{i,j}) = \nabla^{ss} u$$

A.2 Fluid Variables

Navier–Stokes equation: NSE
Computational fluid dynamics: CFD
Partial differential equation: PDE
Density ρ, velocity u, pressure p
First and second viscosities: μ and ξ
Dynamic or shear viscosity: μ
Kinematic viscosity: $\mu/\rho =: \nu$
Reynolds number: $\mathrm{Re} := \frac{\rho VL}{\mu}$
Froude number: $\mathrm{Fr} := \frac{V}{\sqrt{L|g|}}$
Fluid vorticity: $\omega = \nabla \times u$
Cauchy stress vector or traction vector: $\overrightarrow{t}$
Fluid acceleration: $a = \frac{\partial u}{\partial t} + (u \cdot \nabla)u$
Stress tensor: $\Pi = \Pi(x, t)$
Deformation tensor: $\nabla^s u = \mathbf{D} = \frac{1}{2}(\nabla u + \nabla u^t)$

Viscous stress tensor: $\mathbf{V} := \Pi + p\mathbf{I}$
Kinetic energy: $k(t) := \frac{1}{|\Omega|}\frac{\rho_0}{2}\int_\Omega |u(x,t)|^2\,dx$
Energy dissipation rate: $\varepsilon(t) := \frac{\mu}{|\Omega|}\int_\Omega |\nabla u(x,t)|^2\,dx$

A.3 Basic Function Spaces and Norms

Bounded domain (region that encloses the fluid): $\Omega \subset \mathbb{R}^d (d = 2, \text{or } 3)$
Boundary of flow domain: $\partial\Omega$
Outward unit normal to $\partial\Omega$ or to a given subdomain: $\widehat{n}$
Function space $L^2(\Omega)$ denotes the set of all functions $p : \Omega \to \mathbb{R}$ with

$$||p|| := \left[\int_\Omega |p(x)|^2 dx\right]^{1/2} < \infty,$$

$L_0^2(\Omega) := \{q(x) \in L^2(\Omega) : \int_\Omega q(x)dx = 0\}$
Vector valued functions:

$$L^2(\Omega)^d = \{v = (v_1, \ldots, v_d) : \Omega \to \mathbb{R}^d : v_j \in L^2(\Omega), j = 1, \ldots, d\}$$
$$||v|| = ||v||_{L^2(\Omega)^d} := \left[||v_1||^2 + ||v_2||^2 + \cdots + ||v_d||^2\right]^{\frac{1}{2}}$$

Tensor-valued functions:

$$L^2(\Omega)^{d\times d} := \{\mathbf{V} = \mathbf{V}_{ij} \in L^2(\Omega), i, j = 1, \ldots, d\},$$
$$||\mathbf{V}|| := \left[\sum_{i,j=1}^{d} ||\mathbf{V}_{ij}||^2\right]^{\frac{1}{2}}$$

Inner products $(\cdot, \cdot)$ given by

$$(p, q) := \int_\Omega p(x)q(x)dx \text{ for } p \text{ and } q : \Omega \to R,\ p, q \in L^2(\Omega),$$
$$(u, v) := \int_\Omega \sum_{i=1}^{d} u_i v_i dx \text{ for } u, v \in L^2(\Omega)^d,$$
$$(\mathbf{S}, \mathbf{T}) := \int_\Omega \sum_{i,j=1}^{d} \mathbf{S}_{ij}(x)\mathbf{T}_{ij}(x)dx \text{ for } \mathbf{S}, \mathbf{T} \in L^2(\Omega)^{d\times d}$$

A.3.1 Other norms

Norms on L^p spaces:

$$||v||_{L^p} := \left[\int_\Omega |v|^p dx\right]^{1/p},\ 1 \le p < \infty,$$

for example, the $L^4(\Omega)$ norm: $||v||_{L^4} := [\int_\Omega |v|^4 dx]^{1/4}$
The $L^\infty(\Omega)$ norm is the essential supremum over Ω: $||v||_{L^\infty} = \text{ess sup}_{x\in\Omega} |v(x)|$.

A.4 Velocity and Pressure Spaces and Norms

Velocity and pressure spaces (X, Q) given by

$$X := \{v : \Omega \to \mathbb{R}^d : v \in L^2(\Omega), \nabla v \in L^2(\Omega) \text{ and } v = 0 \text{ on } \partial\Omega\},$$

$$Q := \left\{q : \Omega \to \mathbb{R} : q \in L^2(\Omega) \text{ and } \int_\Omega q\, dx = 0\right\} \text{ (also denoted } L_0^2(\Omega))$$

Norms and inner products on X:

$$(u, v)_X := \int_\Omega \nabla u \cdot \nabla v + uvdx \text{ and } ||v||_X = \sqrt{(v, v)_X}$$

Equivalent norm on X: $||\nabla v||$
Dual space of X: $X^* = H^{-1}(\Omega)$ is the closure of $L^2(\Omega)$ in $||\cdot||_{-1}$, where

$$||f||_{-1} := \sup_{v \in X} \frac{(f, v)}{||\nabla v||}$$

Divergence free subspace V of X: $V = \{v \in X : (q, \nabla \cdot v) = 0\ \forall q \in Q\}$
V^* norm: $||f||_* := \sup_{v\in V} \frac{(f,v)}{||\nabla v||}$
Some finite constants:

$$M = M(\Omega) := \sup_{u,v,w \in X} \frac{(u \cdot \nabla v, w)}{||\nabla u||||\nabla v||||\nabla w||},$$

$$N = N(\Omega) := \sup_{u,v,w \in V} \frac{(u \cdot \nabla v, w)}{||\nabla u||||\nabla v||||\nabla w||}$$

Continuous inf-sup condition:

$$\inf_{q\in Q} \sup_{v\in X} \frac{(q, \nabla \cdot v)}{||\nabla v||\,||q||} \geq \beta > 0$$

Discrete inf-sup condition: for β^h bounded away from zero uniformly in h,

$$\inf_{q^h\in Q^h} \sup_{v^h\in X^h} \frac{(q^h, \nabla \cdot v^h)}{||\nabla v^h||\,||q^h||} \geq \beta^h > 0$$

Spaces for time-dependent problems:

$$L^2(0, T; H_0^1(\Omega)) = \left\{v(x, t) : [0, T] \to H_0^1(\Omega) \colon \int_0^T ||\nabla v||^2\, dt < \infty\right\},$$

$$L^\infty(0, T; L^2(\Omega)) = \left\{v(x, t) : [0, T] \to L^2(\Omega) \colon \text{ess}\sup_{0<t<T} ||v|| < \infty\right\},$$

$$L^2(0, T; L_0^2(\Omega)) = \left\{q(x, t) : [0, T] \to L_0^2(\Omega) \colon \int_0^T ||q(t)||^2 dt < \infty\right\},$$

$$L^2(0, T; V) = \left\{v(t) : [0, T] \to V : \int_0^T ||\nabla v||^2\, dt < \infty\right\}.$$

A.5 Finite Element Notation

Finite element method: FEM
Triangulation or finite element mesh: $T^h(\Omega)$
Finite element space on given finite element mesh: $X^h = X^h(T^h(\Omega))$
Nodal interpolant in the chosen finite element space: $(I_h u)(x, y)$
Cubic bubble function associated with the element K: $\phi_K(x, y)$
Velocity, pressure finite element spaces $X^h \subset X$ and $Q^h \subset Q$
Discretely divergence free subspace:

$$V^h := \{v^h \in X^h : (q^h, \nabla \cdot v^h) = 0 \, \forall \, q^h \in Q^h\}$$

Approximate velocity and pressure: $(u^h, p^h) \in (X^h, Q^h)$
MINI element: C^0 piecewise linears enhanced by cubic bubble functions for the velocity and C^0 piecewise linears for the pressure:

$$\begin{aligned} X^h &= \{v^h \in X : v^h|_\Delta \in \mathcal{P}_1(\Delta) \, \forall \Delta \text{ in the mesh}\} \oplus \{\phi_\Delta : \forall \Delta \text{ in the mesh}\}, \\ Q^h &= \{q^h \in L_0^2(\Omega) \cap C^0(\Omega) : q^h|_\Delta \in \mathcal{P}_1(\Delta) \forall \Delta \text{ in the mesh}\} \end{aligned}$$

Explicitly skew symmetrized trilinear form:

$$b^*(u, v, w) := \frac{1}{2}\,(u \cdot \nabla v, w) - \frac{1}{2}(u \cdot \nabla w, v)$$

Finite constants:

$$\begin{aligned} N^h = N^h(\Omega) &:= \sup_{u^h, v^h, w^h \in V^h} \frac{(u^h \cdot \nabla v^h, w^h)}{||\nabla u^h|| ||\nabla v^h|| ||\nabla w^h||} \quad \text{and} \\ M^h = M^h(\Omega) &:= \sup_{u^h, v^h, w^h \in X^h} \frac{(u^h \cdot \nabla v^h, w^h)}{||\nabla u^h|| ||\nabla v^h|| ||\nabla w^h||}. \end{aligned}$$

A.6 Turbulence

Large eddy simulation: LES
Kolmogorov's theory of turbulence developed in 1941: K41 theory
Kolmogorov microscale: $\eta = O(\mathrm{Re}^{-3/4})$ in three dimensions, and $O(\mathrm{Re}^{-1/2})$ in two dimensions
Length scale of filter: δ
Long time averaging: $\langle \phi \rangle(x) := \lim_{T \to \infty} \frac{1}{T} \int_0^T \phi(x, t) dt$
Kinetic energy distribution functions:

$$\begin{aligned} E(k, t) &= \frac{L}{2\pi} \sum_{|\mathbf{k}|_\infty = k} \frac{1}{2} |\widehat{u}(\mathbf{k}, t)|^2 \text{ and} \\ E(k) \, &: = \langle E(k, t) \rangle F \end{aligned}$$

Universal Kolmogorov constant: α around 1.4 to 1.7
Wave number: k

Wave number vector: $\mathbf{k}$
Energy dissipation rate: ε
Units of indicated quantity: $[\cdot]$
Subfilter scale stress tensor (sometimes called the Reynolds stress)

$$\mathbf{R}(u, u) := \overline{u\, u} - \bar{u}\, \bar{u}$$

SFNSE (space-filtered NSE):

$$\bar{u}_t + \nabla \cdot (\bar{u}\, \bar{u}) + \nabla \bar{p} - Re^{-1} \Delta \bar{u} + \nabla \cdot \mathbf{R}(u, u) = \bar{f} \text{ and } \nabla \cdot \bar{u} = 0$$

Turbulent viscosity coefficient: $\nu_T = \nu_T(\delta,\ \bar{u})$
Differential filter: given $\phi(x)$, $\overline{\phi}$ is the solution, under appropriate boundary conditions on $\partial\Omega$, of

$$-\delta^2 \Delta \overline{\phi} + \overline{\phi} = \phi \text{ in } \Omega.$$

Bibliography

[1] M. Ainsworth and J. T. Oden, *A Posteriori Error Estimation in Finite Element Analysis*, Wiley Interscience, New York, 2000.

[2] M. Anitescu, W. Layton and F. Pahlevani, Implicit for local effects and explicit for nonlocal effects is unconditionally stable, *ETNA* 18 (2004), 174–187.

[3] Archimedes (T. L. Heath, transl.), *The Works of Archimedes*, Dover, New York, 2002.

[4] R. Aris, *Vectors, Tensors and the Basic Equations of Fluid Mechanics*, Dover, New York, 1962.

[5] F. Armero and J. C. Simo, Long-term dissipativity of time-stepping algorithms for an abstract evolution equation, *CMAME* 131 (1996), 41–90.

[6] D. Arnold, F. Brezzi, and M. Fortin, A stable finite element for the Stokes problem, *Calcolo*, 21 (1984), 337–344.

[7] O. Axelsson and V.A. Barker, *Finite Element Solution of Boundary Value Problems: Theory and Computation,* SIAM, Philadelphia, 2001.

[8] A. J. Baker, *Finite Element Computational Fluid Mechanics*, Hemisphere, Washington, 1983.

[9] G. A. Baker, *Galerkin Approximations for the Navier–Stokes Equations*, Technical Report, Harvard University, 1976.

[10] G. A. Baker, V. Dougalis, and O. Karakashian, On a higher order accurate, fully discrete Galerkin approximation to the Navier-Stokes equations, *Math. Comp.* 39 (1982), 339–375.

[11] S. Basu, E. Foufoula-Georgiou, and F. Porte-Agel, *Predictability of Atmospheric Boundary Layer Flows as a Function of Scale*, UMn Report, UMSI 2002/89, 2002.

[12] G.K. Batchelor, *An Introduction to Fluid Dynamics*, Cambridge University Press, Cambridge, UK, 2002.

[13] G. K. Batchelor, *The Life and Legacy of G. I. Taylor*, Cambridge University Press, Cambridge, UK, 1996.

[14] L. C. Berselli, T. Iliescu, and W. Layton, *A Mathematical Introduction to Large Eddy Simulation*, Springer, Berlin, 2005.

[15] S. C. Brenner and L. R. Scott, *The Mathematical Theory of Finite Element Methods*, Springer, Berlin, 1994.

[16] F. Brezzi and M. Fortin, *Mixed and Hybrid Finite Element Methods*, Springer-Verlag, New York, 1991.

[17] M. O. Bristeau, R. Glowinski, and J. Periaux, Numerical methods for the Navier-Stokes equations: Applications to the simulation of compressible and incompressible flows, *Comput. Phys. Reports*, 6 (1987), 73–187.

[18] P. Constantin and C. Doering, Energy dissipation in shear driven turbulence, *Phys. Rev. Lett.*, 69 (1992), 1648–1651.

[19] P. Constantin and C. Foias, *Navier-Stokes Equations*, University of Chicago Press, Chicago, 1988.

[20] M. Crouzeix and P.-A. Raviart, Conforming and nonconforming finite element methods for solving the Stokes equations, *R.A.I.R.O.* R3, 7 (1973), 37–76.

[21] T. Dantzig, *Henri Poincaré: Critic of Crisis*, Scribner, New York, 1954.

[22] O. Darrigol, *Worlds of Flow, A History of Hydrodynamics from the Bernoullis to Prandtl*, Oxford University Press, Oxford, UK, 2005.

[23] R. Dautray and J.-L. Lions (I. N. Sneddon, transl.), *Mathematical Analysis and Numerical Methods for Science and Technology: Physical Origins and Classical Methods Volume* 1, Springer-Verlag, New York, 1984.

[24] M. A. Day, The no-slip condition of fluid dynamics, *Erkenntnis* 33 (1990), 285–286.

[25] C. Doering and C. Foias, Energy dissipation in body-forced turbulence, *J. Fluid Mech.* 467 (2002), 289–306.

[26] C. R. Doering and J. D. Gibbon, *Applied Analysis of the Navier-Stokes Equations*, Cambridge University Press, Cambridge, UK, 1995.

[27] Q. Du and M. Gunzburger, Analysis of a Ladyzhenskaya model for incompressible viscous flow, *JMAA* 155 (1991), 21–45.

[28] M. van Dyke, *Album of Fluid Motion*, Parabolic Press, Stanford, 1982.

[29] M. Eckert, *The Dawn of Fluid Dynamics*, Wiley-VCH, Weinheim, Germany, 2006.

[30] J. P. Eckman, *Roads to turbulence in dissipative dynamical systems*, *Rev. Modern Phys.* 53 (1981), 643–654.

[31] H. Elman, D. J. Silvester, and A. J. Wathen, *Finite Elements and Fast Solvers with Applications in Incompressible Fluid Dynamics*, Oxford University Press, Oxford, UK, 2005.

[32] C. L. Fefferman, *Official Clay Prize Problem Description: Existence and Smoothness of the Navier-Stokes Equations*, 2000, http://www.claymath.org/millennium/.

[33] J. H. Ferziger and M. Peric, *Computational Methods for Fluid Dynamics*, 3rd ed., Springer, Berlin, 2001.

[34] P. M. Gresho and R. L. Sani, *Incompressible Flows and the Finite Element Method, volume* 2, John Wiley and Sons, Chichester, UK, 2000.

[35] C. Foias, What do the Navier-Stokes equations tell us about turbulence? *Contemporary Mathematics* 208 (1997), 151–180.

[36] U. Frisch, *Turbulence*, Cambridge University Press, Cambridge, UK, 1995.

[37] G. P. Galdi, *An Introduction to the Mathematical Theory of the Navier-Stokes Equations, Volume* I, Springer, Berlin, 1994.

[38] G. P. Galdi, *An Introduction to the Mathematical Theory of the Navier-Stokes Equations, Volume* II*: Nonlinear Steady Problems*, Springer, Berlin, 1998.

[39] G. P. Galdi, An Introduction to the Navier-Stokes Initial-Boundary Value Problem, 1-70, in *Fundamental Directions in Mathematical Fluid Mechanics*, Birkhäuser, Basel, 2000.

[40] G. P. Galdi, *Lectures in Mathematical Fluid Dynamics*, Birkhäuser-Verlag, Basel, 2000.

[41] G. P. Galdi, Weighted Energy Methods in Fluid Dynamics and Elasticity, *Springer Lecture Notes in Mathematics* 1134, Springer, Berlin, 1985.

[42] G. P. Galdi and W. J. Layton, Approximation of the large eddies in fluid motion II: A model for space-filtered flow, *Math. Models Methods Appl. Sciences* 10 (2000), 343–350.

[43] M. Germano, Differential filters of elliptic type, *Phys. Fluids* 29 (1986), 1757–1758.

[44] V. Girault and P.-A. Raviart, *Finite Element Approximation of the Navier-Stokes Equations*, Springer-Verlag, Berlin, 1979.

[45] V. Girault and P.-A. Raviart , *Finite Element Methods for Navier-Stokes Equations: Theory and Algorithms*, Springer-Verlag, Berlin, 1986.

[46] M. D. Gunzburger, *Finite Element Methods for Viscous Incompressible Flows: A Guide to Theory, Practice, and Algorithms*, Academic Press, New York, 1997.

[47] Y. He, Two-level method based on finite element and Crank–Nicolson extrapolation for the time-dependent Navier–Stokes equations, *SIAM J. Numer. Anal.* 41 (2003), 1263–1285.

[48] J. G. Heywood and R. Rannacher, Finite element approximation of the nonstationary Navier–Stokes problem. I. Regularity of solutions and second-order error estimates for spatial discretization, *SIAM J. Numer. Anal.* 19 (1982), 275–311.

[49] J. G. Heywood and R. Rannacher, Finite element approximation of the nonstationary Navier–Stokes problem. II. Stability of solutions and error estimates uniform in time, *SIAM J. Numer. Anal.* 23 (1986), 750–777.

[50] J. G. Heywood and R. Rannacher, Finite element approximation of the nonstationary Navier–Stokes problem. III. Smoothing property and higher order error estimates for spacial discretization, *SIAM J. Numer. Anal.* 25 (1988), 489–512.

[51] J. G. Heywood and R. Rannacher, Finite element approximation of the nonstationary Navier–Stokes problem. IV. Error analysis for second-order time discretization, *SIAM J. Numer. Anal.* 27 (1990), 353–384.

[52] J. Hoffman, On duality-based a posteriori error estimation in various norms and linear functionals for large eddy simulation, *SIAM J. Sci. Comput.* 26 (2004), 178–195.

[53] J. Hoffman and C. Johnson, *Computational Turbulent Incompressible Flows*, Springer, Berlin, 2007.

[54] T. J. Hughes and J. Marsden, *A Short Course in Fluid Mechanics,* Publish or Perish, Houston, 1976.

[55] T. J. Hughes, A. Oberai, and L. Mazzei, Large eddy simulation of turbulent channel flows by the variational multiscale method, *Physics of Fluids* 13 (2001), 1784–1799.

[56] T. Iliescu and W. Layton, Approximating the larger eddies in fluid motion III: The Boussinesq model for turbulent fluctuations, *Analele Stinfice ale Universitatii " Al. I. Cuza" Iasi* 44 (1998), 245–261.

[57] V. John, Large Eddy Simulation of Turbulent Incompressible Flows. Analytical and Numerical Results for a Class of LES Models, *Lecture Notes in Computational Science and Engineering* 34, Springer-Verlag, Berlin, 2004.

[58] V. John and W. J. Layton, Analysis of numerical errors in large eddy simulation, *SIAM J. Numer. Anal.* 40 (2002), 995–1020.

[59] V. John and W. Layton, Approximating local averages of fluid velocities, Computing 66 (2001), 269–287.

[60] V. John, W. Layton, and N. Sahin, Derivation and analysis of near wall models for channel and recirculating flows, *Computers Mathematics Applications*, 48 (2004), 1135–1151.

[61] C. Johnson, *Numerical Solution of Partial Differential Equations by the Finite Element Method,* Cambridge University Press, Cambridge, UK, 1987.

[62] O. LADYZHENSKAYA, *The Mathematical Theory of Viscous Incompressible Flow*, Gordon and Breach, New York, 1969.

[63] W. LAYTON, A. LABOVSCHII, C.C. MANICA, M. NEDA, AND L. REBHOLZ, *The Stabilized, Extrapolated Trapezoidal-Galerkin Finite Element Method*, Technical Report, Mathematics, University of Pittsburgh, 2006.

[64] J. LERAY, Sur le mouvement d'un fluide visquex emplissant l'espace, *Acta Math.* 63 (1934), 193–248.

[65] J. LERAY, The physical facts and the differential equations, *American Math. Monthly* 61 (1954), 5–7.

[66] M. LESIEUR, *Turbulence in Fluids*, Kluwer Academic Publishers, Norwell, MA, 1997.

[67] D. K. LILLY, The representation of small-scale turbulence in numerical simulation experiments, in *Proceedings IBM Scientific Computing Symposium on Environmental Sciences*, Yorktown Heights, New York, 1967, 195–210.

[68] E. MACCURDY, *The Notebooks of Leonardo daVinci*, George Braziller, New York, 1958.

[69] J. M. MAUBACH, Local bisection refinement for N-simplicial grids generated by reflection, *SIAM J. Sci. Comput.* 16 (1994), 210–227.

[70] J. C. MAXWELL, On the condition to be satisfied by a gas at the surface of a solid body, *Scientific Papers* 2 (1879), 704.

[71] R. E. MEYER, *Introduction to Mathematical Fluid Dynamics*, Dover, New York, 1982.

[72] B. MOHAMMADI AND O. PIRONNEAU, *Analysis of the k-epsilon Turbulence Model*, Wiley, New York, 1994.

[73] P. ORLANDI, *Fluid Flow Phenomena: A Numerical Toolkit*, Kluwer, Boston, 1999.

[74] O. PIRONNEAU, *The Finite Element Methods for Fluids*, Wiley, Chichester, UK, 1983.

[75] H. POINCARÉ, *Theorie des Tourbillons (1893)*, Editions Jacques Gabay, Paris, 1990.

[76] S. POPE, *Turbulent Flows,* Cambridge University Press, Cambridge, UK, 2000.

[77] G. PRODI, Un teorema di unicita per le equazioni di Navier-Stokes, *Ann. Mat. Pura Appl.* 48 (1959), 173–182.

[78] L. QUARTAPELLE, *Numerical Solution of the Incompressible Navier-Stokes Equations*, Birkhäuser Verlag, Basel, 1993.

[79] L. REBHOLZ, *An Energy and Helicity Conserving Finite Element Scheme for the Navier-Stokes Equations*, Technical Report, Mathematics, University of Pittsburgh, 2006.

[80] O. REYNOLDS, On the dynamic theory of incompressible viscous fluids and the determination of the criterium, *Phil. Trans. R. Soc. London A* 186 (1895), 123–164.

[81] L. F. RICHARDSON, *Weather Prediction by Numerical Process*, Cambridge University Press, Cambridge, UK, 1922.

[82] N. ROTT, Note on the history of the Reynolds number, *Annual Review of Fluid Mechanics* 22 (1990), 1–11.

[83] S. G. SADDOUGHI AND S. V. VEERAVALLI, Local isotropy in turbulent boundary layers at high Reynolds number, *J. Fluid Mech.* 268 (1994), 333–372.

[84] P. SAGAUT, *Large Eddy Simulation for Incompressible Flows*, 2nd ed., Springer-Verlag, Berlin, 2002.

[85] H. M. SCHEY, *Div, Grad, Curl, and All That: An Informal Text on Vector Calculus*, 3rd ed., W.W. Norton, New York, 1996.

[86] H. T. SCHLICHTING, K. GERSTEN, E. KRAUSE, H. JR. OERTEL, AND C. MAYES, *Boundary Layer Theory*, 8th ed., Springer, Berlin, 2004.

[87] J. SERRIN, Mathematical principles of classical fluid mechanics, *Handbuch der Physik*, Springer, Berlin, 1959, 125–263.

[88] J. SERRIN, The initial value problem for the Navier-Stokes equations, *Nonlinear Problems*, University of Wisconsin Press, Madison, 1963, 69–98.

[89] J. C. SIMO, F. ARMERO, AND C. A. TAYLOR, Stable and time-dissipative finite element methods for incompressible Navier-Stokes equations in advection dominated flows, *IJNME* 39 (1995), 1475–1506.

[90] J.S. SMAGORINSKY, General circulation experiments with the primitive equations, *Mon. Weather Rev.* 91 (1963), 99–164.

[91] K. R. SREENIVASAN, On the scaling of the turbulent energy dissipation rate, *Phys. Fluids* 27 (1984), 1048–1051.

[92] K. STEWARTSON, D'Alembert's paradox, *SIAM Rev.* 23 (1981), 308–343.

[93] G. STRANG AND G. J. FIX, *An Analysis of the Finite Element Method*, Prentice–Hall, Englewood Cliffs, NJ, 1973.

[94] B. STRAUGHAN, *The Energy Method, Stability and Nonlinear Convection*, Springer, Berlin, 2003.

[95] R. TEMAM, *Navier-Stokes Equations: Theory and Numerical Analysis*, AMS, Providence, RI, 2001.

[96] F.-C. THOMASSET, *Implementation of Finite Element Methods for Navier-Stokes Equations*, Springer-Verlag, Berlin, 1981.

[97] G. A. TOKATY, *A History and Philosophy of Fluid Mechanics*, Henley-on-Thames, Foulis, 1971.

[98] C. TRUESDELL, *Essays in the History of Mechanics*, Springer, Berlin, 1968.

[99] S. TUREK, S. KILIAN, AND S. L. TUREK, *The Virtual Album of Fluid Motion*, Springer-Verlag, Berlin, 2002.

[100] R. VERFÜRTH, *A Posteriori Error Estimation and Adaptive Mesh Refinement Techniques*, Wiley, New York, 1996.

[101] X. WANG, The time averaged energy dissipation rates for shear flows, *Physica D* 99 (1997), 555–563.

[102] G. B. WHITHAM, *Linear and Nonlinear Waves*, Wiley-Interscience, New York, 1974.

Index

L^2, 5

asymptotic stability, 132

backward Euler, 152, 153, 164, 177, 196
basis functions, 22
bilinearity, 28
body force, 73
Boltzmann equation, 72
boundary layer, 87, 88, 177
Boussinesq assumption, 186

Cauchy stress principle, 93
Cauchy stress vector, 73–75, 82, 97
Cauchy–Schwarz inequality, 7
Cea's lemma, 31, 35, 68, 124
checkerboard mode, 60
coating flows, 81
coercive, 29
compact support, 11, 142, 187
computational fluid dynamics (CFD), xiii, 18, 66, 79, 161, 163, 164, 175, 195
conservation, 43
  of mass, 72
  of momentum, 44, 45, 48, 50, 73
conservation of mass, 72
continuity equation, 44, 66, 89
contraction, 108, 125
  tensor, 38
convection-diffusion equation, 45
cross product, 39
Crouzeix–Raviart element, 66
cubic bubble, 25, 64
curl, 40, 42, 49, 67, 96, 98
cutoff length scale, 186

d'Alembert's paradox, 119
Darcy's law, 47
deformation, 37, 39, 76
differential filter, 196
dimensional analysis, 85, 180
direct numerical simulation (DNS), 180
discrete inf-sup condition
  Ladyzhenskaya–Babuska–Brezzi
    (LBB) condition
  div-stability, 62, 66, 131, 135
discretely divergence free
  Vh, 62
div, 40
divergence free subspace V, 57
divergence theorem, 26, 42, 72
driven cavity, 79
dual norm, 105
dynamic similarity, 13, 85
dynamic viscosity, 76

eddy viscosity, 186, 188
energy cascade, 14, 186
energy dissipation rate, 146
energy equality, 148
energy inequality, 139, 143, 145, 146
Euler equations, 12, 76, 87
extrapolated trapezoid rule, 163–165

Fick's law, 45
finite element method (FEM), 26, 30, 32, 33, 35, 54, 60, 62, 69, 132, 151, 165, 168, 175, 177

Fortin's lemma, 63
Fourier's law of heat conduction, 46
Froude number, 98
fully discrete, 151
fundamental problem, 143

Galerkin approximation, 30, 125
  Stokes problem, 59
Galerkin orthogonality, 61
Gaussian filter, 196
global uniqueness, 107, 125
grad, 40
gradient tensor, 39
Gronwall's inequality, 148, 152

Hölder's inequality, 7, 117
Helmholtz decomposition theorem, 43
Hood–Taylor element, 66, 131, 161
hyperviscosity, 195

incompressible, 42
inf-sup condition, 57, 58, 63, 67–70, 104, 117, 199
interpolation, 23
irrotational flow, 42, 50
isolated solution, 116
iteration, 134

K41 theory, 14, 182
kinematic viscosity, 78
kinetic energy, 3, 5, 139, 145
kinetic theory, 78
Kolmogorov microscale, 180, 185

Ladyzhenskaya inequalities, 11
Ladyzhenskaya model, 194
large eddy simulation (LES), 180, 189–191, 194
Lax–Milgram theorem, 28, 34, 35, 58, 67
LBBh, 122, 126, 154
  discrete inf-sup condition, 63
Leray conjecture, 146
Leray theory, 180
Leray–Prodi–Serrin condition, 144, 149
Leray–Schauder, 111, 112
linear-constant element, 59

little lemma, 60
Lp, 7

material derivative, 50, 74, 93
MINI element, 64, 66, 130, 160
mixed methods, 54
momentum equation, 75

Navier's slip law, 82
Navier–Stokes equations (NSE), 72, 77, 78, 85, 88, 89, 97
Newton's method, 119
Newtonian fluid, 77
no-slip condition, 8, 78, 79
nonsingular solution, 116
nonsingularity, 117

Oseen problem, 108, 119, 176

passive scalar, 45
penalty method, 69
perfect fluid, 76
Poincaré–Friedrichs inequality, 9, 15, 28, 57, 105, 106, 132, 195
porosity, 46
porous media, 26
potential flow, 42, 50, 53
pressure, 12, 17, 26
pressure drag, 99
pressure Poisson equation, 17, 18, 28, 68
Prodi–Serrin condition, 144, 149

regularity conjecture, 146
Reillich lemma, 111
Reynolds averaging, 97
Reynolds number, 13, 83, 85, 132, 151, 177, 189
Reynolds stress tensor, 188

semidiscrete, 151
Serrin's condition, 144
shear flow, 40, 110
shear viscosity, *see* dynamic viscosity
skew symmetric, 133
slip with friction, 78, 193, 196
Smagorinsky model, 191
space-filtered NSE, 188

spin tensor, 39
Stokes flow, 86
Stokes problem, 53–64, 66–69
Stokes projection, 176, 177
stress, 74
stress strain relation, 76
stress tensor, 75, 76, 94, 97, 148, 194
strong solution, 140, 148, 153
summation convention, 38
support of a function, 65, 142

Taylor cells, 91, 115
Taylor experiment, 91, 114
time average, 149
Trace theorem, 10
traction vector, 73
trapezoid rule, 161
triangulation, 18
turbulence, 151, 180
turbulent pressure, 189
turbulent stress, 188
turbulent viscosity, 189

uniqueness, 143

V* norm, 57, 105, 117
variational formulation, 28, 55–58, 108, 140
    Stokes problem, 56
viscosity, 12
viscous stress tensor, 76
von Karman vortex street, 87, 101
vorticity, 40, 96, 98

weak derivative, 10
weak solution, 139, 141, 142

X, 8
X*, 30
X* norm, 9